Electronics texts for engineers and scientists

Editors

H. Ahmed, *Reader in Microelectronics*, *Cavendish Laboratory*,
University of Cambridge

P. J. Spreadbury, *Lecturer in Engineering*, *University of Cambridge*

An introduction to control and measurement with
microcomputers

Electronic texts for engineers and scientists

C. Oatley
Electric and magnetic fields

H. Ahmed & P. J. Spreadbury
Analogue and digital electronics for engineers

K. F. Sander & G. A. L. Reed
Transmission and propagation of electromagnetic waves, second edition

R. L. Dalglish
An introduction to control and measurement with microcomputers

AN INTRODUCTION TO

control and measurement
with microcomputers

R.L.DALGLISH

School of Physics, University of New South Wales

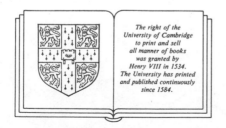

*The right of the
University of Cambridge
to print and sell
all manner of books
was granted by
Henry VIII in 1534.
The University has printed
and published continuously
since 1584.*

CAMBRIDGE UNIVERSITY PRESS

Cambridge

New York   New Rochelle   Melbourne   Sydney

Published by the Press Syndicate of the University of Cambridge
The Pitt Building, Trumpington Street, Cambridge CB2 1RP
32 East 57th Street, New York, NY 10022, USA
10 Stamford Road, Oakleigh, Melbourne 3166, Australia

First published 1987

Printed in Great Britain by Scotprint, Musselburgh, Scotland

British Library Cataloguing in Publication Data
Dalglish, R. L.
An introduction to control and measurement
with microcomputers.–(Electronics texts
for engineers and scientists)
1. Automatic control–Data processing
2. Microcomputers
I. Title   II. Series
629.8'312   TL223.M53

Library of Congress Cataloguing in Publication Data
Dalglish, R. L. (Robert Louis), 1936–
An introduction to control and measurement with
microcomputers.

(Electronics texts for engineers and scientists)
Includes bibliographies and index.
1. Microcomputers. 2. Computer interfaces. 3. Automatic data collection systems. I. Title. II. Series.
TK7888.3.D293 1987   502.8'5416   86-30970

ISBN 0 521 30175 0 hard covers
ISBN 0 521 31771 1 paperback

Contents

Introduction

This book can be used as a guide to personal study, as a text book for formal study or by anyone who just wants to know how a computer really works. It is based on courses given in the School of Physics at the University of New South Wales since 1980. These courses serve several aims. They provide Science students with the basic knowledge needed to use microcomputers as scientific instruments, they provide Computer Science students with an understanding of computer hardware and, in the form of an evening course given to a wide spectrum of the general public, they provide a sound knowledge of how computers work.

The emphasis is not on the conventional 'software' approach to computing. Instead, it is on the internal operation of the computer hardware and the internal structure or 'architecture' of the machine. This is intended to provide a mechanistic understanding of software operation, and to allow a full exploitation of a microcomputer in a range of applications wider than can be achieved from the conventional software approach. This expands the use of the machine to research data collection or experiment control and to 'real-time' process control.

The role of computers has changed considerably since their introduction in the late 1940s. The first machines were designed to perform the tedious, repetitive calculations needed to explore numerical models of Nature which were not solvable by analytic techniques. They were thus used first by scientists and engineers for numerical modelling of real-world systems. These early machines were difficult to use, expensive and very limited. As the machines grew in complexity and storage capacity, techniques were developed to simplify their use. This made them suitable for commercial data processing and storage. The size of this market thenceforth guided the development of the computer industry towards commercially oriented machines. The combination of this prime commercial market and the technology available at that time, favoured the development of very powerful but expensive centralised units. The speed and capacity of these *mainframe* machines allowed them to service many terminals at the same time; they were *time-shared*.

The separation of the user from these remote machines, where communication is usually by a punched card reader or a remote keyboard terminal, allows the user only to run batched or time-shared programs; this evokes little interest in how the machine actually works. In the majority of computer applications it is therefore necessary for the user to understand only the machine's *terminal characteristics*, that is, how it appears to operate from the user's terminal. Since the user seldom if ever interacts directly with the central computer hardware, there is no incentive to understand it.

Prior to the development of *integrated circuit semiconductor devices*, these mainframe, time-shared machines were the only economical solution for powerful computers. Since direct access to the central hardware was not possible, the formal study of the theory of computing, *Computer Science*, concentrated on the user's view of the machine, that is, the formal study of programming (*Software* and *Software Systems*). The other aspect of computing, *hardware* (the electronic and mechanical components of the computer system), has only been of interest to computer engineers.

There are some applications where mainframe machines are used in *real-time interactive* applications, that is, where the computer and its system software are used to monitor continuously and control technical or industrial processes, but the cost of mainframe machines limits such applications to large production plants or to situations such as military applications where none of the costs are considered.

The recent development of *Large Scale Integrated* (LSI) and *Very Large Scale Integrated* (VLSI) silicon-chip technology has dramatically changed this situation and has made available the full power of the computer in a very cheap form. This has changed dramatically the scope of computer applications. The principal product of this new technology, the personal microcomputer, is fully accessible at the hardware level. It has therefore the potential to be not only an interactive software machine, but also to be a powerful instrument for data collection or control. This opens a wide field of new applications for the computer.

But here is the problem! The formal study and teaching of computing has not yet caught up with the explosive development of the microcomputer. Thus the full capability of the personal microcomputer as a data-collection or control instrument has not yet been realised; it is available only to those who are lucky enough to have an interest in computers *and* a good knowledge of electronics.

Such a wasteful situation cannot be tolerated, especially since contemporary microcomputers are simple enough to be understood adequately with very little knowledge of electronics and only a little more about computing. With this minimal knowledge, it is possible to become master of the machine and to make it do your bidding.

This book is intended to provide the essential basic information. It assumes only a knowledge of basic electricity and some experience in writing and entering simple programs in a high-level language such as *BASIC*, *PASCAL* or *FORTRAN*. The material is presented in a carefully designed progressive manner such that the *general principles* become clear. Only that information which is necessary for this understanding is presented here, but the method of presentation is intended to give the reader the basic knowledge needed to pursue any of the topics. Each chapter ends with some exercises (where applicable) and some suggestions for further reading.

The text describes the internal operation of computers in general terms. The 6502 microprocessor is used for detailed examples because it is very simple and easy to understand, and is used in two convenient microcomputers, the Apple and the BBC. The first six chapters describe the fundamentals of a computer's internal operation; the last three chapters describe how it is expanded into a powerful system and how it is interfaced to the outside world.

The courses on which this book is based involve an integrated program of laboratory work using simple electronic equipment. These are replaced in this book by descriptive segments interspersed within the text and by worked examples. Details of the demonstration equipment are available from the author to anyone interested in developing similar courses.

1

Computers and binary logic

1.1   A brief history

The internal structure and organisation of contemporary computers can be best understood when one has some knowledge of their evolution. We begin therefore with a very brief look at the history of computers.

The need for computing machines came originally from Science and Mathematics. The role of Science is to attempt to understand natural processes by devising and developing testable 'models' of Nature. In the study of Physics, the models are usually presented as mathematical relationships which predict the variation of some prescribed parameters of the natural system. Unfortunately, these relationships can seldom be handled analytically; that is to say that they cannot be manipulated directly into the form of an *explicit* equation in observable or testable parameters.

Newton's fundamental laws of motion can be manipulated in such a way that the entire history of interaction between two gravitational (non-relativistic) rigid bodies can be represented analytically as direct equations describing their motions in time and space. However, if a third rigid body is introduced, the same laws *cannot* be manipulated to give similar explicit equations of motion.

These *non-analytic* equations dominate Science and have generated a large field of study in Mathematics devoted to solution techniques for these problems. *Numerical* techniques involving simple but repetitive calculations can generally be used to solve the equations. The technique is used extensively to produce tables of logarithms and trigonometric functions. These tables are crucial for navigation; in the Nineteenth Century the Royal Navy's astronomical and mathematical tables contained many errors. In 1812, Charles Babbage realised that the errors could be eliminated if the tabulations were calculated *and typeset* by an automatic machine. Techniques had been devised in the Eighteenth Century to calculate these tables by sequences of simple additions and subtractions. In 1784, the French mathematician Baron de Prony, charged with the unenviable task of converting all French trigonometric tables to a decimalised angular quadrant of 100°, had developed a system based on the *finite differences* technique. A large number of unskilled workers were instructed in simple tasks involving accepting numbers, processing them and passing the results on to others. Careful design of this arrangement allowed complicated calculations to be carried out cheaply and quickly. Babbage set about the *mechanisation* of such a process. in 1822 a proposal for such a machine, the *Difference Engine*, was accepted by the British government. Work began on its construction with the intention of producing accurate, error-free navigation tables for the Royal Navy.

Box 1.1
Finite differences

A very simple description is outlined here, without the underlying theory, of the finite difference technique.

Consider the function

$$y = x^2$$

A tabulation of the values of the dependent variable y for integer values of the independent variable x is required.

To evaluate the $(x+1)$th value, we consider the two previous values,

$$x^2, \quad (x-1)^2 = x^2 - 2x + 1$$

the difference between the two previous values (the first difference)

$$x^2 - (x-1)^2 = 2x - 1$$

and the differences between these first differences (the second difference)

$$(2x-1) - [2(x-1) - 1] = 2$$

Since the second differences are constant, the third differences will be zero.

The value of any square can be expressed as the sum of the previous square and the last two differences

$$(x+1)^2 = x^2 + 2x + 1$$
$$= x^2 + (2x-1) + 2$$

Obviously the entire table of squares can be generated from the first three entries by an extremely simple algorithm involving only addition.

A table of cubes is generated in the same manner from the first four cubes by adding the previous cube to the last three differences. This is left as an exercise.

This process is very suitable for automatic execution in a machine since it requires only very simple arithmetic operations and is controlled entirely by the initial values set into the machine. The algorithm can be generalised from integer to fractional step size and can be extended to generate tabulations of any constant coefficient polynomial of a single variable. Since most single variable functions can be expressed as constant coefficient polynomial series, the technique therefore can be used to generate tabulations of virtually any function. If the first few values of the function are calculated by hand (to adequate precision!), the initial data entries can be loaded into the machine. From there on it is simply a matter of 'turning the handle'.

Babbage's Difference Engine was designed to handle 20-digit decimal numbers and sixth-degree polynomial functions. The final register and the difference registers were each connected by adder mechanisms. When the machine was run, the final (result) register automatically generated successive values of the function determined by the initial settings of all the registers; this value was automatically transferred to soft metal plates which were then used as moulds for printing plates.

The project went well for about four years but disagreements between Babbage, the government and the construction engineers terminated it before the machine was complete. Two Swedes, Georg and Edvard Deutz, later built a simpler machine from Babbage's publications. It was less sophisticated and handled 15 digits for third-degree polynomials. It was brought to Britain in 1855 and operated there for a few years, but its limitations and unreliability soon led to its abandonment.

While developing the Difference Engine, Babbage realised that a machine could be made to carry out a much greater variety of numerical calculations if its sequence of operations was not limited to the simple fixed requirements of finite difference calculations, but could instead be varied or *programmed* at will. In about 1834, he began development of this more powerful machine called the *Analytic Engine*, the first programmable computer. He worked intermittently on its development for the rest of his life.

In his original design, the variable sequence of operations was determined by pinned barrels similar to those used in a music box. As the machine's design evolved, he replaced the barrels with the card system developed in France between the Mid-eighteenth Century and 1805 for the Jacquard weaving looms. Babbage gave the machine the ability to vary its sequence of operations conditional on the sign of the number in its main register; he also provided a means for the sequence cards to be stepped backwards and forwards, so that a particular sequence of operations could be repeated. The provision of these two features gave the machine enormous computing power. L. F. Menabrea and Ada Augusta Byron (Lady Lovelace) were very impressed by its latent power, and wrote articles and papers on its flexibility, describing many of the essential features of modern 'programming'.

Box 1.2
The analytic engine layout

The Analytic Engine consisted of two principal parts, the 'store' and the 'mill' (Fig. 1.1). The store was simply a memory array which was intended eventually to hold 1000 50-digit numbers, any of which could be accessed as required. The mill controlled the sequence of operations to be carried out under the command of the 'master barrel' which, in the first design, was programmed by the insertion of pins, to carry out the required sequence. This master barrel could call numbers from the store and could invoke numerical operation sequences defined as 'subroutines' on 'subsidiary barrels' on the numbers in the store. When Babbage replaced the barrels with Jacquard punch cards, he immediately realised that the machine would then be able to punch out intermediate values for later use. It would then be able to generate its own data and could, given enough time, carry out calculations of any degree of complexity.

Figure 1.1   The layout of the original Analytic Engine.

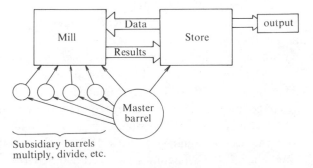

Subsidiary barrels
multiply, divide, etc.

Babbage developed the concepts of the Analytic Engine faster than they could be realised as hardware. This caused major problems with its construction. Like the Difference Engine, this remarkable machine was never completed. If either of his machines had been brought into operation, the growth of computing and its effect on the structure of our society would have been much earlier and, perhaps, very different.

Babbage's concepts were too far ahead of their time. The notion of a programmable machine was replaced during the following 80 years by the much simpler concepts of mechanical calculators. These were extensively developed in the early Twentieth Century. Two developments in the concept of programmable machines appeared, however, in the 1930s. Konrad Zuse, working in Germany between 1930 and 1941 and unaware of Babbage's work, developed a series of programmable calculators using electromechanical switching relays. From 1936 on, the Cambridge mathematician Alan Turing published papers describing a logical 'thought' machine which could be used to demonstrate that there is no way of determining whether or not a particular mathematical statement is provable! Apart from this esoteric application in the philosophy of mathematics, Turing described how such a machine could be instructed to carry out any logically definable task as an automatic sequence of simple logic operations. He applied this concept to a series of electromechanical and electronic machines leading to the *Colossus* used to carry out routine code-breaking at Bletchley Park during World War II. He continued to develop his concepts, and in 1946 he proposed the *Automatic Computing Engine* (*ACE*). This project foundered amidst bureaucratic bungling and austerity in postwar Britain. Meanwhile, in the USA, Howard Aiken of Harvard University, with the co-operation of the International Business Machines Corporation (IBM), developed a machine named the *Automatic Sequence Controlled Calculator* (*ASCC*) which first ran in 1944. It seems that Aiken was aware of Babbage's work; the ASCC bears many resemblances to the layout of the Analytic Engine. It was entirely mechanical, but was controlled by a punched paper tape carrying instructions and input data.

An electronic machine, the *Electronic Numerical Integrator And Calculator* (*ENIAC*) was being developed concurrently by J. W. Mauchly and J. P. Eckert at the University of Pennsylvania. The prime purpose of this machine was to compute artillery ballistics tables for the US Army. It had 20 registers each of 10 decimal digits, employed some 18,000 thermionic vacuum tubes and weighed 30 tons. It first ran in 1946.

The ASCC and the ENIAC machines were both based on decimal number registers. Mechanical instruments can be designed to operate reliably in any number base, since each digit of the number can be represented by a shaft which is indexed into the required number of positions. All of the base-10 calculating machines which have been available for many years use this principle, as do the base-12 and base-60 clocks which have been marking our time for the last 500 years.

Electronic machines, however, are better suited to operation in base-2 where the possible states of an element are reduced to two, OFF or ON (see Box 1.3). The ENIAC machine overcame the decimal problem by using arrays of complex electronic decimal (ten-state) ring counters. Unknown to these workers, the machines made by Zuse had used binary integers with 'floating-point' arithmetic (more about this in Chapter 3).

Box 1.3
Decimal or binary?

Mechanical counters can be designed to work in any number base. The base-10 odometer in a motor car is quite reliable and seldom makes mistakes unless assisted by used-car salesmen. Decimal adder mechanisms are also quite reliable. Babbage's mills and the later mechanical calculators were all sophisticated examples of such mechanisms. These machines are reliable because mechanical 'noise' is easily controlled; random fluctuations in the system are much smaller than the 'interval' between adjacent 'states' of the mechanism. Each state is defined by an indexed position of the counter wheels. It takes quite a jolt to make the counters of an odometer change from one state to another. (Used-car salesmen would deny this.)

Electromechanical systems, where the indexed mechanisms are operated by solenoid or relay actuators, can also be very reliable. Early automatic telephone exchanges used such mechanisms.

Serious problems arise, however, with electronic systems. Here the 'state' of a variable must be defined by either a voltage or a current. Both of these quantities are subject to electromagnetic induced noise. It is, furthermore, much more difficult to establish multiple, well-defined states of a voltage or current variable (see Box 1.4).

Therefore, for simple, reliable, electronic systems, the number of states of a variable should be restricted to two. This was realised in the earliest days of telegraphy when the two states were simply whether or not current was flowing in the telegraph wire (Fig. 1.2).

The technology of telegraphy rapidly developed into the formal study of binary logic control systems (more of this in Section 1.6) using electromechanical relays (Fig. 1.3). These are simple devices where the magnetic field produced by a current flowing in a solenoid causes an iron block to actuate a switch. These can have only two states but, over a wide range of solenoid current, the states are well defined. In other words, this simple binary device has a high immunity to 'noise'. Similarly, simple electronic devices function most reliably with only two defined states.

The ENIAC machine used decimal integers each defined by ten, two-state variables organised in ten-state 'rings'. These rings were arranged such that only one member could be on at a time. Ten elements can represent 1024 separate states if the 'one-only' restriction is removed. This highly inefficient use of electronics was recognised and avoided in subsequent 'binary' machines.

Figure 1.2   An early telegraph 'detector' as described by Prof. J. Henry in 1832.

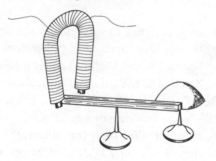

Figure 1.3   An early repeater relay as described by Prof. J. Henry in 1832.

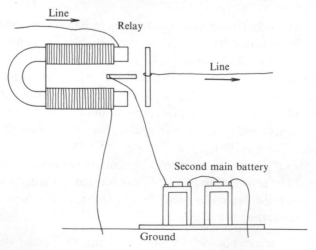

The *Electronic Discrete Variable Computer (EDVAC)* conceived by John von Neumann in about 1945 used binary logic, thermionic valves, mercury delay lines and magnetic wire memory. It was the first machine which merged its program with data in the main memory. It was the first of the modern computers and led to the many machines built between the late 1940s and the early 1950s, collectively referred to as *first generation computers*. Classical logic had been developed and extended to study and understand binary logic and switching systems. This knowledge was easily transferred to binary machines and allowed arithmetic operations to be expressed in terms of simple, binary logic algorithms. The *logic* machines that resulted from this were not limited to arithmetic applications and thus became more powerful, general-purpose data processors.

The invention of the transistor in 1948, and its application to computer technology, marked the next major step in computer development. These transistor-based *second generation* machines were much more compact, more powerful and more economic than the first-generation leviathans. Prior to this, computers had been used almost exclusively by scientists and engineers. These new general-purpose machines were well suited to a wide range of applications in commerce and technology. Competition for this enormous market brought about extremely rapid development. The use of transistors allowed faster, more compact machines but still required a lot of design and assembly time. Because of this, the most economical configuration continued to be very large centralised machines operated by specialists remote from the programmer. These large second-generation mainframe computers dominated computer development for many years.

From about 1965 the development of *integrated circuits*, where many semiconducting devices and their ancillary resistors etc. are all integrated onto a single chip of silicon, made the circuitry required for an electronic computer very compact and allowed the development of the *third generation* machines and the first compact computer, the *minicomputer*. These *minis* were the first computers small enough, yet adequately powerful, to be brought into the computer user's own workplace. They were still quite expensive and were not very powerful by present standards, but they allowed direct interaction between the machine and the user. They made direct data collection and process control possible.

The rate of development since the introduction of VLSI components and the resulting *fourth generation* machines has been meteoric.

Computer Science courses, with a necessary lead time of several years, have been unable to keep up with the explosive development of the prime product of VLSI technology, the *personal microcomputer*. There is little doubt that it is this very class of machine that will have the most profound effect on Society in general and on Science and Technology in particular. Let me substantiate this prediction.

Before the development of integrated circuits, the cost of computers was so great that they could only be employed in those situations where they introduced adequate economies (commercial applications), where there were no viable alternatives (research) or where cost was not considered (defence). The formal study of computer applications, *Computer Science*, evolved within this environment to provide suitable programmers, engineers and technicians.

The resulting formalism, oriented predominantly towards software (programming), directed subsequent developments in hardware and led to the application of Computer Science and Technology in the fields of *Process Control*. The requirement there is to design systems to collect data, process and analyse it, and then to adjust process control parameters to optimise the behaviour of the system. If this control loop is established by a programmable logic machine (a specialised computer), two desirable properties result. Firstly, if the process is changed or modified, optimisation of the new system can be established with only a change to the system program. Secondly, and more significantly, it is then unnecessary to design and develop new hardware for each new system that is needed; the same hardware can be used over and over again for different applications with new programs.

This rationalised concept inevitably led to the introduction of a general-purpose integrated logic processor based on standard computer architecture and with a suitable set of inbuilt logic operations (instructions) or *microprogram*. It was soon realised that this *microprocessor* could be used as the core of a small compact computer. The microprocessor was immediately adapted to perform this new role and started the *microcomputer* revolution.

To return to the prediction made above, it has not yet been fully realised that the microcomputer brings the power, economy and flexibility of programmed logic process control and data collection to the computer user in a cheap, convenient and easily accessed form. A new approach to computers, based principally on their internal hardware operation, is needed to exploit their full potential properly.

The purpose of this book is to provide the basic knowledge needed for this in a friendly and easily understood manner. The logical first step is to explore the quintessence of the computer's operation, *binary logic*.

1.2   Binary logic, variables and operations

At the most fundamental level, a computer is simply a machine which carries out *logic operations* on *logic variables*. We must therefore first explore the notion of logic and operations on logic variables.

The formal study of logic originated as a series of treatises based on lecture courses given by Aristotle on systematic reasoning in about 340 BC; they were later collected under the title of the *Organon*. The statements or propositions of Aristotle's logic can only have one of two possible values, true or false; they are *binary logic variables*, the very core of a computer's internal operation. The aspect of logic with which we are concerned here is the study of the validity of arguments based on common language propositions or statements. The original purpose was to develop theorems about the truth or falsity of compound statements, based on the truth or falsity of their elements. This well-established formalism is at the very core of the design and operation of a computer.

We will begin by introducing the concepts of binary logic variables in the historical, traditional manner by way of common language statements.

Consider the statement

All quirts are black

Let us label this statement with the logic variable symbol **A**. The truth value (true or false) of the statement **A** cannot be established until a quirt is defined, but the truth value of the derived statement (we will call it **B**),

Some quirts are black

is dependent on the truth value of **A**. If **A** is true, then **B** will also be true. If **A** is false, no conclusion can be inferred about **B**. However, if **B** is false, then **A** is also false and so on.

The formal study of such rules of *inference* for complicated chains of argument has fascinated logicians since the time of Aristotle as the 'Traditional Formal Logic'. We are not concerned in this book with the rules of inference. We only need the concept of binary logic variables and will move on to another aspect of logic, the effect of combining logic statements.

Now consider the statement

It is evening

At some times in a day this will be true, at others it will be false. Obviously it is a binary logic variable. Let us label it again as **A**.

Here is another statement which we will label **B**.

Rain is falling

Both of these logical variables are independent in that the value of either one has no effect on the other. We can generate another dependent statement by operating on the original two with the common language conjunction '*and*',

It is evening and rain is falling

Let us label this new dependent logic variable **C**. Then, in symbols

$$C = A \quad \text{and} \quad B$$

From our understanding of common language, we can derive the truth value of **C** from the values of **A** and **B**.

Let us, for convenience, assign 'truth' the numerical value '1' and 'falsity' the value '0'. Each of the independent variables can be '1' or '0'. Thus there are 2×2 (four) possible combinations. We will now draw up a *truth table* to show the value of **C** in terms of the independent variable values. If all four combinations are depicted we will have exhausted all possible conditions for the *operation* of 'and' between **A** and **B** (Truth Table 1.1). Since this truth table exhausts all possible conditions it fully defines the 'and' operation between two variables. It is therefore a better definition of the operation than the rather vague notion that our common language assigns to the conjunction.

Truth Table 1.1

A	B	C
1	1	1
1	0	0
0	1	0
0	0	0

From here on we will use this truth table as the formal definition of the 'and' operation and represent it by the label '**AND**'. Since the table looks like multiplication a dot ($\cdot$) is conventionally used to symbolise it. The symbolic equation

$$A \cdot B = C$$

now represents the truth table above and fully defines the operation

$$A \cdot B$$

By considering the common language meaning of the conjunction '*or*', a truth table can also be drawn up to define formally the 'or' operation. A small problem arises here because of a possible ambiguity about the truth value of the statement '**A or B**' when both **A** and **B** are true. Does this mean that the compound statement is true or false? We can resolve this ambiguity simply by defining the operation as shown in Truth Table 1.2.

Truth Table 1.2

A	B	C
1	1	1
1	0	1
0	1	1
0	0	0

This '**OR**' operation is symbolised by the $+$ sign, so that the truth table above is represented by the expression

$$A + B = C$$

and fully defines the operation

$$A + B$$

The ambiguous common interpretation of the first row of the table below is included into the formalism by defining another operation which expresses the alternate interpretation (Truth Table 1.3). This operation is called the '*exclusive-or*' (**EOR**) and represents the common language operation inferred by 'one *or* the other, but not both'. The operation is symbolised thus

$$A \oplus B$$

Truth Table 1.3

A	B	C
1	1	0
1	0	1
0	1	1
0	0	0

Before we look into the theorems that can be developed from these simple definitions, there is one other common language operation that we must formalise. This is the operation characterised by the phrase 'it is not true that ...' If we precede any statement with this phrase, the truth value of the compound statement becomes opposite to that of the original. We can define this relationship, which is called '*negation*', by Truth Table 1.4. This is symbolised by a horizontal bar

$$B = \bar{A}$$

Truth Table 1.4

A	B
0	1
1	0

We are not going to use the formalism to study the inferences of common language. Since we only wish to use it to study binary logic and operations, we can now discard all reference to its common language source.

Compound operations can be defined by combining the 'primitives' explored above. For instance, the operation obtained by negating the result of the **AND** (defined by Truth Table 1.5) is called the **NAND** operation, symbolised

$$\overline{A \cdot B}$$

Truth Table 1.5

A	B	A·B	$\overline{A \cdot B}$
1	1	1	0
1	0	0	1
0	1	0	1
0	0	0	1

and the operation obtained by negating the **OR** (Truth Table 1.6) is called the **NOR** operation and is symbolised

$$\overline{A + B}$$

Truth Table 1.6

A	B	$\overline{A + B}$
1	1	0
1	0	0
0	1	0
0	0	1

(These last two operations will be used extensively later in this chapter. Note that the **NAND** operation gives a *true* result if any variable is *false* and that the **NOR** operation gives a *false* result if any variable is *true*.)

1.3   Binary (digital) electronic logic

The concept of binary variables discussed in Section 1.2 can be modelled with electric and electronic systems if the two allowed states of the variables are defined as the presence or absence of a voltage or current. Voltages and currents are of course continuous variables and could in principle be used to specify an arbitrary number of different states of a variable, but they are susceptible to drift and noise which limit the number of states that can be reliably defined. (See Box 1.4.)

The most efficient logic system is that which has only two states. This binary or *digital* logic system was used in early telegraphy, where it was only necessary to detect the presence or absence of a current flowing in the telegraph line. From this early and economically important beginning, the application of binary logic has grown into a powerful field of knowledge about digital switching systems. (See Box 1.5.)

Box 1.4
Noise and drift

Noise in a system is unwanted signals which degrade the true information signals. In the early telegraph, the main source of noise was the loss of signal strength over long distances. The telegraph line introduced resistance, capacitance and inductance to impede or degrade the actual signal. The solution for long lines was to place repeaters at distances over which the signal could be reliably detected. These were simple electromechanical relays which used the feeble incoming signal current to close contacts which forced current from a local battery into the next section of the line (Fig. 1.3). This simple technique could only be used with a binary signal.

The situation is a little more complicated in electronic systems. In most digital (i.e. binary) electronic systems, the signal variable is a voltage. Two voltage states must be defined, one for logic '0' and one for logic '1'. The signal lines are bathed in electromagnetic radiation from other signal lines and from external sources. The devices have non-zero impedance and the radiation thus superimposes noise signals on the real logic signal.

The signal output from any simple practical device will change slightly (drift) with age, load or temperature change. The actual '0' and '1' logic variables are thus subject to several sources of noise; the two states of the logic variable must be defined therefore, as ranges rather than simple levels. A circuit built from such devices will only function correctly if the devices are designed and connected to reject all noise and to respond only to the real logic signal. This is achieved by setting the '**1**' and '**0**' signal voltage ranges far enough apart that they cannot ever be corrupted by the effect of noise or drift. The difference between these logic levels, making allowances for manufacturing variations, determines the noise margin for the logic system. The expected noise level in the system must be less than this noise margin.

Box 1.5
Switching systems

Electromechanical relays model binary logic. Relays are simple devices which use an electric current to energise an electromagnet which in turn attracts an iron armature connected to one or more electric switches. The switches can be *Normally Open* (*NO*) (the energising current will close them), or *Normally Closed* (*NC*) (the energising current will cause them to open). Fig. 1.4 depicts a simple relay. The simplified diagram of Fig. 1.5 is generally used in systems with many relays.

Relays can model binary logic operations. Fig. 1.6 shows realisations of the **AND**, **NAND**, **OR** and **NOR** operations. Arbitrarily complicated operations can be implemented by relay circuits. Before the invention of the transistor, such circuits were the most economical way of expressing logic operations for automatic control systems. The high cost of building and maintaining such circuits made it crucial that the simplest possible

Figure 1.4   A schematic representation of an electromechanical relay.

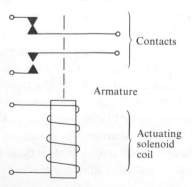

Figure 1.5   The simplified symbolic representation of a relay.

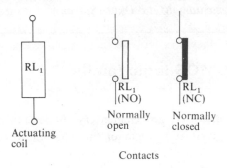

Figure 1.6   A segment of a relay system to generate several
operations.

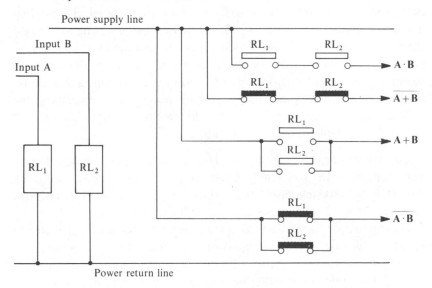

realisation of a required logic function be found. A lot of theoretical work
was carried out on *minimisation* of logic functions.

When binary computers were conceived, the vast body of knowledge
about switching systems was immediately applied to their design. The
development of electronic logic devices stimulated the rapid development
of modern computers, and the commercial competition to build faster and
more compact computers caused rapid evolution of the electronic devices
used. While this evolution is very interesting, it is not essential to an
understanding of the computer hardware. We will therefore go directly to

the present standard 'families' of devices, *TTL* (*Transistor–Transistor Logic*) and *CMOS* (*Complementary Metal Oxide Silicon*).

Box 1.6
TTL, LS and CMOS logic families

Note: this box assumes an elementary understanding of transistors; references are given at the end of this chapter to suitable sources of information.

Many families of electronic logic devices have been developed since the mid-1950s. Several different types of logic operation called gates are available in any *logic family*. We will only discuss briefly the families currently used in microcomputers. An understanding of the internal operation of the logic devices is not essential to their use; one can regard them simply as *black boxes* which, by some incomprehensible magic, perform their tasks. In this book, however, where we are developing a sound knowledge of microcomputer hardware operation, a brief look inside the black box is in order. The main reason is to understand the restraints and limitations that must be observed when they are used. Full details are available in the cited literature.

The most common device family in current use is *Low-power Schottky* (*LS*), a development of the TTL family. Other variations of the family are L (Low power), S (Schottky clamped) and F (Fast). All are very similar and, with care, are compatible. We look first at the TTL logic family.

The internal structure of a typical three-input **NAND** gate is depicted in Fig. 1.7. This discloses some limitations of a real TTL logic device.

    (1) The outputs can 'sink' a limited amount of current to 'ground'.

    (2) The inputs must either be held at an adequately high voltage or must have some current drawn from them at an adequately low voltage.

    (3) If the outputs are to be used as the input for subsequent gates, the output levels must equal or exceed the requirements of the next input.

    (4) A change at the input to the gate will require that all of the transistors change their state and that the parasitic capacitance of the output line be charged; some finite time is thus required for a change to propagate through to the output.

    (5) If two logic outputs are connected together, then, if they endeavour to signal different output levels, the result cannot be properly defined. (We will consider this further in a later chapter.)

The first four limitations are described in terms of *logic thresholds*, *noise margins*, *fan-out* and *propagation delay*. The logic thresholds are simply the limits defined for the two states of the variable. To allow for safe, error-free chaining, i.e. connecting one gate's output to drive a following gate's input, the output logic voltage levels are designed to be within limits which are more stringent than those for the input voltages. The difference between these thresholds determines the device's immunity to noise. Since each input requires current to be sunk by its preceding gate, there is obviously a limit to how many inputs can be connected to any output. The manufacturer guarantees that the input and output levels are maintained under specified load conditions. This in turn determines the fan-out of the gates in the family.

Threshold levels are shown in Fig. 1.8; under worst state conditions, there will always exist a margin which ensures that a subsequent input

Figure 1.7   Schematic diagram of a three-input TTL **NAND** gate. The gate on the left has all three inputs at logic **true** (high). Since the input voltages are above the threshold determined by R_2, R_3, TR_1 and TR_2, TR_3 is cut-off and current flows into the base of TR_4.

TR_4 thus 'sinks' current from subsequent gates as shown in the gate on the right; TR_2 is cut-off and the bias current from R_2 flows into the base of TR_3. This allows current from R_4 to 'pull up' the output voltage Q.

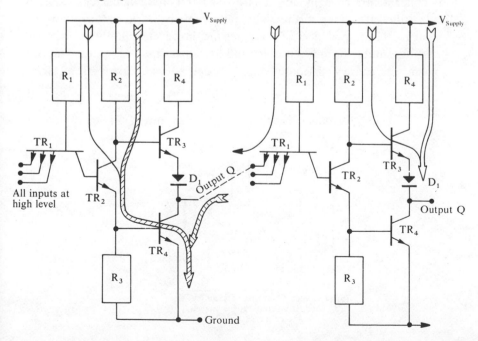

always interprets the prior output correctly. Provided that no noise disturbance exceeds this margin, no errors will result. This is called the noise margin. The gates themselves are a prime source of noise since, when they change state, a current surge is impressed on the supply line. Care must be taken to minimise these surges. The time taken to change state is defined by the propagation delay for the two possible transitions. For TTL, these are about 11 ns (low to high) and 7 ns (high to low).

TTL is now almost universally replaced by the LS family. The logic levels are almost the same, but the currents are quite different. For TTL the nominal input current is 1.6 mA and the rated output current is 16 mA, giving a maximum fan-out of 10. For LS the currents are 0.36 mA and 8 mA, giving a fan-out of about 20. The lower currents of LS would imply a higher impedance and thus a slower, more noise-sensitive system; this is avoided by using Schottky transistors to yield wider noise margins and faster transitions. Both families have fairly high power consumption. A typical 'package' of gates, a set of four two-input **NAND** gates draws about 12 mA (TTL) or 2.5 mA (LS).

Another common family, CMOS, allows low-power logic systems (Fig. 1.9). The CMOS gates are inherently simpler since the transistors act like ideal complementary switches and thus need no resistors. In either of their two states, virtually no current is drawn from the supply. The only significant current is, in fact, the transient and charging current required when the gate changes state. This favours their use in battery powered systems. The impedances of the gates are, however, quite high; they are very susceptible to induced noise and are slower than LS gates. The two voltage states are well separated and the noise margins are made as big as

Figure 1.8   Guaranteed input and output ranges (TTL and LS).

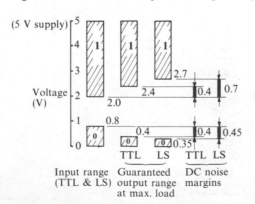

possible to counteract the noise sensitivity. The gates can operate with supply voltages from 3 to 15 V.

Very high speed gates are available in another family called *Emitter Coupled Logic* (*ECL*). These devices use unsaturated difference amplifiers which respond very quickly and have no supply transients. Propagation delays are about 2 nS per gate. They are used in very fast logic processors.

Logic gates are usually supplied in the industry standard *Dual-In-Line* (DIL) plastic or ceramic package (see Fig. 1.10).

Figure 1.9　(*a*) A simplified schematic of a CMOS gate (two-input **NOR**). (*b*) Noise margins (for a 5 V supply).

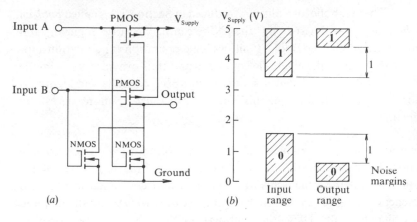

(*a*)　　(*b*)

Figure 1.10　A typical DIL plastic package.

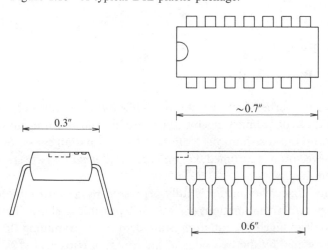

Operations on logic variables (Section 1.2) can be easily *modelled* in simple, cheap, integrated circuit, semiconductor logic gates. In the early days of semiconductor devices, when the gates were quite expensive, the manufacturers attempted to gain economies of scale by minimising the number of different types of gates in a family. The theory of binary logic shows how this can be achieved; the next section explores this as a small sample of the power of digital (binary) logic, and to set the stage for an understanding of a microcomputer's internal operation.

1.4   Boolean algebra

The symbolic formalism introduced in Section 1.2 is called *boolean algebra*; it was devised by the Nineteenth Century logician and mathematician George Boole. It allows logic operations and relationships to be manipulated like algebraic equations and thus provides a systematic means to explore and discover subtle or unexpected results and identities. Before indulging in some of this, the algebra must be more rigorously defined.

The notion of logic variables and the truth tables for logic operations in Section 1.2 represent the basis of boolean algebra. A full set of the postulates and their theorems are presented and developed in the cited references; we simply state them here since we only wish to use them, not to study them formally. Using the symbols from Section 1.2, some properties of boolean algebra are expressed below.

(1) $A + B = B + A$

(2) $A \cdot B = B \cdot A$

(3) $A \cdot (B + C) = A \cdot B + A \cdot C$

These three properties demonstrate the similarity of the rules of boolean algebra to those of ordinary algebra. They are like the algebraic notions of commutation and distribution. Note that the normal algebraic order of operation execution is assumed, i.e. brackets (and *negation*) are strongest, '·' next and finally '+'.

Since any logic relationship is fully expressed by a truth table which depicts all possible combinations of states, it is possible to show whether or not any boolean algebraic equation is an identity by examining the truth tables represented by the two sides of the equation (Box 1.7).

Box 1.7
Proofs of theorems by truth table

In the example below, a truth table representation is used to prove the property expressed by Equation (3).

Since there are three independent variables there are eight possible combinations or states. To exhaust all of these, eight rows are needed.

As a first step, draw up the truth table for the 'innermost' operation of the LHS. The next operation expressed by the LHS of the equation can then be performed by consulting the first and fourth column of Truth Table 1.7. The RHS can be treated in the same way to generate, in two stages, the truth table defined by it (Truth Table 1.8). By examining the truth table it is seen that the entries for the LHS (column 5) are in every case the same as those in the RHS (column 8). This proves by exhaustion of all possible states, that the two sides of the equation are identical.

Truth Table 1.7

A	B	C	$(B+C)$	$A \cdot (B+C)$
0	0	0	0	0
0	0	1	1	0
0	1	0	1	0
0	1	1	1	0
1	0	0	0	0
1	0	1	1	1
1	1	0	1	1
1	1	1	1	1

Truth Table 1.8

A	B	C	$(B+C)$	$A \cdot (B+C)$	$A \cdot B$	$A \cdot C$	$A \cdot B + A \cdot C$
0	0	0	0	0	0	0	0
0	0	1	1	0	0	0	0
0	1	0	1	0	0	0	0
0	1	1	1	0	0	0	0
1	0	0	0	0	0	0	0
1	0	1	1	1	0	1	1
1	1	0	1	1	1	0	1
1	1	1	1	1	1	1	1

Here are some further properties or identities.

(4)　$A \cdot 0 = 0$

(5)　$A \cdot 1 = A$

(6)　$A + 0 = A$

(7)　$A + 1 = 1$

All of these are easily proven by the use of truth tables.

Next, consider the mathematical notion of *association*. We postulate that it holds

$$(A \cdot B) \cdot C = A \cdot (B \cdot C)$$

We could decide to express either side thus

$$A \cdot B \cdot C$$

This previously undefined three-variable operation can now be represented completely by its truth table, which can be generated easily by repeated application of the truth table defined for two variables (Truth Table 1.9).

Consider next the following three interesting properties of the *negation* operation. They are all easily proven by truth tables.

(8)　$A + \bar{A} = 1$

(9)　$A \cdot \bar{A} = 0$

(10)　$\overline{(\bar{A})} = A$

Two very useful identities of the algebra are represented by the two expressions of de Morgan's theorem.

(11)　$\overline{A \cdot B} = \bar{A} + \bar{B}$

(12)　$\overline{A + B} = \bar{A} \cdot \bar{B}$

A proof of these is left as an exercise. (Note well that $\bar{A} \cdot \bar{B}$ is not the same as $\overline{A \cdot B}$.) The identities below are important for algebraic manipulations; they

Truth Table 1.9

A	B	C	$A \cdot B$	$(A \cdot B) \cdot C$
1	1	1	1	1
1	0	1	0	0
0	1	1	0	0
0	0	1	0	0
1	1	0	1	0
1	0	0	0	0
0	1	0	0	0
0	0	0	0	0

are easy to prove

(13) $A \cdot A = A + A = A$

Now we consider again the **EOR** operation explored in Section 1.2. It can be expressed in terms of other operations

$$A \oplus B = (A + B) \cdot \overline{(A \cdot B)}$$

This can be proven by the truth table method and is left as an exercise. Although we decided to ignore common language application of logic, note that the RHS of the expression above could be expressed as 'either **A** or **B** or both, but not both **A** and **B**'. The truth table for **EOR** can be represented by yet another logic expression

$$A \oplus B = (A \cdot \overline{B}) + (\overline{A} \cdot B)$$

In the exercises at the end of this chapter, the reader will discover that all the logic operations defined above are intimately related; there are no basic or primitive operations.

Boolean algebra represents a powerful method for representing and manipulating operations on binary variables; Box 1.8 gives two examples. The next section of this chapter outlines an alternative formalism which has provided a general technique for minimisation of expressions.

Box 1.8
Manipulations with boolean algebra

This box shows how boolean algebra can be used to manipulate and simplify logic expressions. Such simplification or minimisation was very important when logic devices were very expensive. Nowadays, the low cost of logic gates allows inefficient logic design. There are, however, still some cases where minimisation is needed.

Let us first gather the boolean logic expressions discussed before

(1) $A + B = B + A$

(2) $A \cdot B = B \cdot A$

(3) $A \cdot (B + C) = A \cdot B + A \cdot C$

(4) $A \cdot 0 = 0$

(5) $A \cdot 1 = A$

(6) $A + 0 = A$

(7) $A + 1 = 1$

(8) $A + \bar{A} = 1$

(9) $A \cdot \bar{A} = 0$

(10) $\overline{(\bar{A})} = A$

(11) $\overline{A \cdot B} = \bar{A} + \bar{B}$

(12) $\overline{A + B} = \bar{A} \cdot \bar{B}$

(13) $A \cdot A = A + A = A$

Example 1.1

Prove that the two alternative expressions given for the **EOR** operation are equivalent. In algebraic terms, prove that

$$(A + B) \cdot \overline{(A \cdot B)} = A \cdot \bar{B} + \bar{A} \cdot B$$

Proof: manipulate the LHS

$$(A + B) \cdot \overline{(A \cdot B)} = (A + B) \cdot (\bar{A} + \bar{B}) \qquad \text{de Morgan} \quad (11)$$
$$= (A \cdot \bar{A} + A \cdot \bar{B}) + (B \cdot \bar{A} + B \cdot \bar{B}) \qquad \text{dist.} \quad (3)$$
$$= 0 + A \cdot \bar{B} + B \cdot \bar{A} + 0 \qquad \text{by} \quad (9)$$
$$= A \cdot \bar{B} + \bar{A} \cdot B \qquad \text{comm.} \quad (2)$$

Example 1.2

Simplify the compound logical statement

$$A \cdot (C + \bar{B}) + A \cdot \bar{C} + \overline{B + A}$$

This expression can be manipulated and simplified thus

$$A \cdot (C + \bar{B}) + A \cdot \bar{C} + \overline{B + A}$$
$$= A \cdot C + A \cdot \bar{B} + A \cdot \bar{C} + \bar{B} \cdot \bar{A} \qquad (3), (12)$$
$$= A \cdot C + A \cdot \bar{C} + \bar{B} \cdot A + \bar{B} \cdot \bar{A} \qquad \text{comm.}$$
$$= A \cdot (C + \bar{C}) + \bar{B} \cdot (A + \bar{A}) \qquad \text{dist.} \quad (3)$$
$$= A + \bar{B} \qquad (9)$$

Both of these examples could be handled by truth tables; direct algebraic manipulation of expressions is generally more efficient for manipulation of complicated expressions and for simplification of moderately complicated expressions.

1.5   Venn diagrams and minimal expressions

Box 1.8 showed how a logic expression could be simplified by boolean algebra. The procedure needs intuition, experience and luck; if the final expression is complicated, one cannot be certain that it is in its simplest or minimal form. This is fine for intellectual exercise, but is inadequate for routine minimisation of engineering or control problems of the type to be outlined in Section 1.6. A systematic approach which allows determination of the minimal expression has been developed over the years. It is based on an alternative graphical model of binary logic operations proposed by John Venn in 1880. In this model the 'space' wherein all statements exist is represented by a bounded surface (a piece of paper!). This space can be divided into two areas; within one area the statement **A** is true and in the other it is false (Fig. 1.11(*a*)). This expresses the boolean identity

$$A + \bar{A} = 1$$

Multiple variables can be represented; Fig. 1.11(*b*) depicts all possible conditions for two logic variables. Within this simple diagram the areas wherein any logic operation on the two variables is true can be depicted. This is shown in Fig. 1.12.

Venn diagrams are a convenient means to visualise logic functions. They can give insight into how a moderately complicated relationship of a few variables can be minimised. The simple graphical representation is very

Figure 1.11 (*a*) The Venn diagram representation of variable **A**. The outer boundary represents the Universe wherein all logic variables exist; the inner area is that region where **A** is **true**. (*b*) All conditions for two variables.

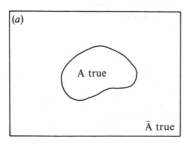

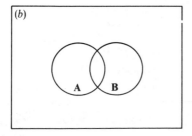

clumsy for expressions with many variables. The basic model has been refined by devising different procedures for *mapping* complicated multi-variable expressions onto a compact logic space diagram. The technique, called *Karnaugh Mapping*, divides the space representing n variables into 2^n rectangles in a systematic manner. This is demonstrated for Example 1.2 of Box 1.8 in Fig. 1.13.

Note that the first step in Example 1.2 was to expand the expression into a 'sum' of 'product' terms. Any boolean expression can be represented thus (the *minterm* expression), or as a product of sums (the *maxterm* expression). When the conditions represented by the minterms are entered onto this

Figure 1.12   Logic operations on Venn diagrams. The 'truth' areas for various operations are shown shaded.

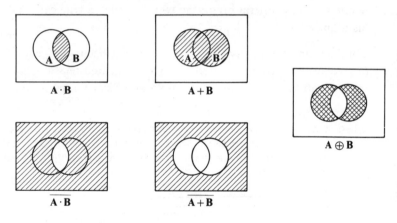

Figure 1.13   Karnaugh Map representation of the expression in Example 1.2 (Box 1.8). The areas where each of the minterms is true are shown on the map. The simplest alternative form of the expression is shown by the pattern of the true terms to be **A + B**.

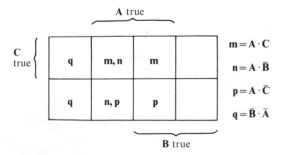

map, the patterns generated can be used to derive a minimal form of the expression.

The Karnaugh Map technique allows minimisation of logic expressions by a systematic graphical procedure. A further development of this principle converts it into a fully defined logical algorithm working on a *tabulation* of the expression. This refinement, the *Quine–McCluskey method*, allows the process to be executed automatically by a computer program.

With the development of cheap, compact microcircuits, the dramatic reduction in the cost of logic devices has reduced the importance of minimisation. It is often more efficient to use standard low-cost, high-level function blocks even where they do not represent the simplest logic expression.

A further brief mention of this topic will be made in Section 8.3; the references at the end of this chapter provide a starting point for further study.

1.6 Logic functions and digital control circuits

We have now seen that boolean algebra allows complicated functions of many logical variables to be modelled and manipulated. We have also found that the logic operations involved in these functions can be expressed with digital electronic logic gates. Box 1.6 showed that these gates can be 'chained' or 'cascaded' to any degree of complexity and can function without logic errors; digital electronic logic can thus be made to model properly any function expressible as a boolean algebraic statement or equation. In this section we will explore a little of the way in which this is done.

Boolean algebra is obviously a convenient way to write down and manipulate logic functions but it is not convenient for expressing the electrical interconnection of the individual gates which model the function. *Schematic diagrams* showing logic signal connections between graphical symbols representing *gates* are invariably used to depict the digital electronic model. Since the development of digital logic, there have been many conventions used for the schematic logic diagrams; the most common of these, MIL-STD-806, is used throughout this book (Box 1.9). The International Electronic Commission (IEC) is developing a new symbol convention to express the complex logic functions now available in LSI and VLSI chips (see references at the end of this chapter).

Box 1.9
Logic symbol conventions

MIL-STD-806 is the most common convention in current use for graphical logic symbols. It defines standard symbol shapes to represent the **AND**, **OR** and **EOR** gates (Fig. 1.14). A further symbol is defined to represent a BUFFER gate; this single input gate does not affect the logic value of the variable, instead, as its name implies, it boosts the logic signal to increase 'fan-out' or changes the logic variable voltage levels to allow different logic families to be interconnected. (The convention defines several other symbols, we will describe these when we need them.)

A *state indicator* is specified as a small circle at inputs or outputs. The effect of this indicator is to negate the logic signal going into, or coming out from, the gate.

Any number of inputs can be represented, but, if there are more than four, the input should be extended as shown in Fig. 1.15.

Figure 1.14   Graphical logic symbols. The shapes define the logic operations on input signals arriving at the 'blunt' end. The result is output at the 'sharp' end.

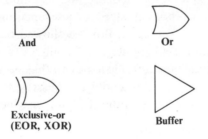

And

Or

Exclusive-or
(EOR, XOR)

Buffer

Figure 1.15   Logic connections on diagrams.

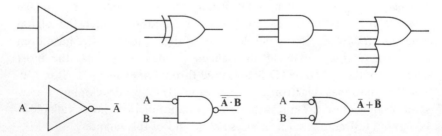

A ———$\bar{A}$

A, B ———$\overline{A \cdot B}$

A, B ———$\bar{A} + \bar{B}$

The standard convention for digital control logic truth tables uses the symbols **H** and **L** for the two possible states of a variable. This convention is used because the *true* state can be defined to be either the higher or the lower voltage. The first of these is called *positive* logic, the latter is *negative* logic. In this book we will always use *positive* logic and will violate the convention by always assigning the higher voltage state the logic value '**1**', and the lower voltage state to '**0**'. (This is the common usage with microcomputer logic circuits.)

Consider the boolean equation

$$\overline{A \cdot B} + C \cdot D + \overline{\overline{E} \cdot \overline{F}} = G$$

This is modelled exactly by the symbolic schematic diagram in Fig. 1.16. Note that the order of logic evaluation implied in the equation determines the connection pattern; the highest order processing of the inputs is done at the first gates encountered by the incoming logic signals.

The principal use for digital electronic logic is in *digital control systems*. In the course of this book we will show that a computer is only a sophisticated digital control system. It is therefore necessary to understand the concept of digital control. We will start with a simple example, and will develop the theme in subsequent chapters.

A control system is used to monitor and adjust the behaviour of some 'natural' or 'real' process so that it is maintained in a desired state or responds in a predetermined manner to random changes. (The words 'natural' and 'real' here refer to a process in Nature. As you will see, this is to differentiate it from the 'model' of the process within the control system.)

The essential elements of a control system are shown in Fig. 1.17. Sensor probes are attached to the natural system to collect and pass information about its *state* to the control system. The control system uses the

Figure 1.16   A schematic diagram representing a logic equation.

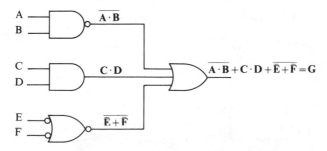

information (data) to generate control signals which are able to modify the behaviour of the natural system. The control system must have an internal model of the natural system; this allows it to interpret the input data and to use its control outputs in such a way that the natural system is forced into some desired state or pattern. An oil refinery is an example; a long, involved, multi-stage process produces a wide range of chemical products from a single feedstock with a variable chemical composition. The purpose of the control system is to monitor the entire process and to adjust the many flow rates, temperatures, etc., so that the most efficient use is made of the feedstock while the desired proportions of the required products are maintained. This example is a little beyond our analysis at this stage. We will examine instead a slightly simpler artificial system. Since we are concerned here with digital systems, we will use an example where only binary inputs and outputs are required.

Consider a bank which requires an alarm system. There is only one output required, namely to sound an alarm under certain defined conditions (this is intended to modify the natural process taking place). Several inputs are required to collect data from the real system and to decide whether an alarm is needed. The layout of the bank is shown in Fig. 1.18. The security officer has decided that the conditions under which an alarm is to be sounded are,

(1) during bank hours if
   (a) the 'panic' alarm switch has been activated, or
   (b) the vault door has been opened without the guard's switch actuated,

   or
(2) outside bank hours if
   (a) the panic alarm switch has been activated, or
   (b) the ultrasonic detector in the vault room detects motion, or
   (c) the vault door is opened, or
   (d) the bank doors are opened without the manager's keyswitch actuated.

Figure 1.17   The basic elements of a control system.

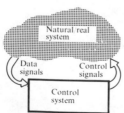

All of these conditions can be written as a boolean expression in binary variable inputs. Let

     P be true if the panic alarm is actuated,

     V be true if the vault door is opened,

     G be true if the guard's switch is actuated,

     U be true if the ultrasonic detector is actuated,

     D be true if the bank doors are open,

     M be true when the manager's keyswitch is actuated, and let

     H be true during bank hours.

The condition for an alarm signal **A** can then be written

$$A = H \cdot P + H \cdot (V \cdot \bar{G}) + \bar{H} \cdot P + \bar{H} \cdot U + \bar{H} \cdot V + \bar{H} \cdot D \cdot \bar{M}$$

This boolean equation can be simplified a little by inspection

$$A = (H + \bar{H}) \cdot P + H \cdot V \cdot \bar{G} + \bar{H} \cdot (V + U + D \cdot \bar{M})$$
$$= P + H \cdot V \cdot \bar{G} + \bar{H} \cdot (V + U + D \cdot \bar{M})$$

and can then be modelled by the symbolic logic diagram of Fig. 1.19.

There is a major difference (other than complexity) between the oil-refinery example and the alarm control depicted above. The alarm system simply responds to the *combination* of binary inputs presented to it at any instant; it cannot modify its response for different *sequences* of data inputs. These are called *combinational logic* systems. The oil refinery control system (and a computer) must respond to the time sequence of events in its real system. Such systems are usually called *sequential logic* systems to discriminate them from combinatorial systems; we will discuss aspects of them relevant to computers in the next chapter.

Figure 1.18   The layout of the bank.

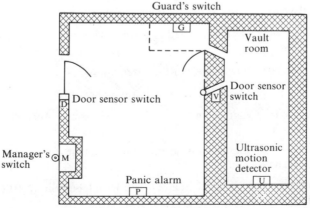

Figure 1.19   The schematic logic diagram of the alarm control system.

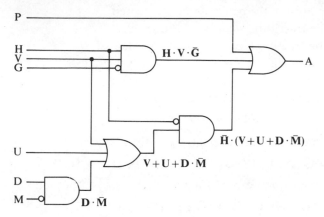

Exercises

1.1 (*a*) Extend the procedure described in Box 1.1 to show that a table of cubes can be generated from the cubes of 1, 2, 3 and 4.

(*b*) Use the technique of finite differences to evaluate the values of the expression

$$y = 6x^3 + 5x^2 + 2x + 11$$

for $x = 0, 1, \ldots, 12$.

(*c*) Show that the technique of finite differences can be extended to arbitrary, uniform intervals of the independent variable.

1.2 Prove the validity of both expressions of de Morgan's theorem using the method of truth tables.

1.3 Show by using truth tables that the **EOR** operation can be represented by the following expression

$$A \oplus B = (A + B) \cdot \overline{(A \cdot B)}$$

1.4 Do the two exercises above by the method of Venn diagrams.

1.5 The following was encountered in a recent Government departmental questionnaire.

Tick the box below if it is not the case that

either you were born before 1964 but not after 1970,

or at least one of your parents were either European or Australian but not British

or you have held your present position for more than 2 years but not for more than 18 months

and furthermore, it is not true that you have found this question difficult to answer.

If you were born in 1963, your father was Dutch and your mother was a North American Indian, you had emigrated 1 year ago and gone straight into a job in the department and you were reasonably intelligent, what would you have done? (Hint: use boolean algebra!)

1.6 This exercise is in the form of a tutorial. It is best performed on digital logic demonstration equipment. Such equipment typically contains an array of digital logic gates whose inputs and outputs are connected to pins on a mimic panel. The various inputs and outputs can be connected to establish logic circuits. Switches are also fitted to allow logic input variables into the circuit, and some logic level detectors, usually small lights allow digital variable outputs to be observed. If there is no such equipment available, you will have to use your imagination.

It has already been mentioned that, in the early days of electronic logic, the manufacturers made the minimum reasonable range of different types of gates. In the first part of this exercise it will be shown that either **NAND** or **NOR** gates alone can generate all of the operations discussed so far. In this exercise **NAND** gates alone will be used.

(*a*) (i) Draw up the truth table for a two-input **NAND** gate and from it derive the table that would result if both inputs are 'tied' together. (Both will always have the same logic value.) Note that the resulting truth table is that of *negation*. Thus the symbols below are equivalent

(ii) Show that the diagram below depicts the **AND** operation

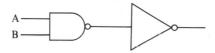

(iii) Show by truth tables that the two diagrams below are equivalent.

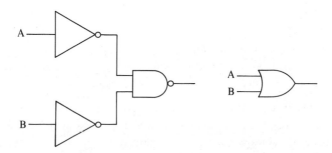

(iv) Extend the diagram of Part (*a*)(iii) to achieve the **NOR** operation.
(v) Determine by inspection, or by truth table, the operation defined
by the diagram below.

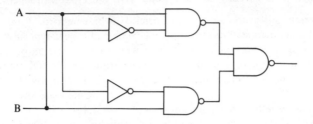

(vi) If you need more practice, generate all the operations above using
NOR gates only.
(*b*) We will soon be concerned with logic 'black boxes' called *decoders*.
These are designed to process the input variables and to generate
separate discrete output signals, each corresponding to a specific input
pattern. There are four possible input patterns for a two-input
decoder. Devise a logic 'black box' which will signal each of the four
possible input patterns with a *low* logic signal at one of four separate
outputs. The diagram below shows the required truth table and a
black-box depiction of it.

Inputs		Outputs			
A	B	C	D	E	F
0	0	0	1	1	1
0	1	1	0	1	1
1	0	1	1	0	1
1	1	1	1	1	0

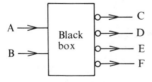

1.7 If you have access to a digital logic demonstration set, draw up
diagrams for the original and simplified expressions in Example 1.2
(Box 1.8). Verify that the behaviour of both is identical.

1.8 If you like inverted logic, explore the result of changing from *positive*
to *negative* logic convention (Box 1.9). Start by looking at the effect
on the truth tables of the simple operations.

References and further reading

These references expand the subject matter discussed in this chapter. The list is not exhaustive, and the cited works may not be those best suited to your level of knowledge.

1 *Charles Babbage*, A. Hyman, Princeton University Press, Princeton (1982)
2 'The Birth of the Computer', *New Scientist*, **99**, No. 1375, 15 Sept. 1984, pp. 778 91
3 *High Speed Computing*, S. H. Hollingdale, English University Press, London (1959)
4 *Computer Architecture and Organisation*, J. P. Hayes, McGraw-Hill, New York (1978)
5 'Colossus: Godfather of the Computer', B. Randell, *New Scientist*, **73**, No. 1038, 10 Feb. 1977, pp. 346–8
6 'Turing Machines', J. Hopcroft, *Scientific American*, May 1984
   References 1–6 give an account of early computers.
7 *The Electric Telegraph – an Historical Anthology*, G. Shiers (ed.), Arno Press, New York (1977)
   A collection of the earliest papers on telegraphy.
8 *The Traditional Formal Logic*, W. A. Sinclair, Methuen, London (1937)
9 *Symbolic Logic*, I. M. Copi, Macmillan, New York (1965)
   References 8 and 9 cover the philosophy of logic.
10 *Introduction to Switching Theory and Logical Design*, F. J. Hill & G. R. Peterson (2nd edition), Wiley, New York (1974)
   This book expands on the application and minimisation of digital switching systems.
11 *Microelectronics*, J. Millman, McGraw-Hill, New York (1979)
12 *The Art of Electronics*, P. Horowitz & W. Hill, Cambridge University Press, Cambridge (1980)
   References 11 and 12 give simple explanations of electronic devices and logic gates.
13 *An Introduction to Numerical Methods and Optimization Techniques*, R. W. Daniels, North-Holland, New York (1978)
   Reference 13 is a good introduction to numerical methods.

Data manuals from the manufacturers of logic chips contain a lot of important information. Texas Instruments TTL Logic manuals describe the new symbolic standard mentioned in Box 1.9.

2

Sequential logic

Preamble

Chapter 1 introduced binary logic but only *combinatorial logic*, where the result of a logic operation depends only on the contemporary state of logic variable inputs, was explored. A logic machine such as a computer obviously needs a wider capability; its behaviour must respond to prior states of its inputs. To be capable of this, it must store information or data and access it during subsequent operations. It also must be able to carry out a sequence of operations and, to be a *universal* machine, there must be provision for modifying the sequence. We will consider the implementation of modifiable sequences in later chapters; in this chapter and the next we will explore the ways in which the simple concepts developed in Chapter 1 for *combinatorial logic* can be extended to provide some of the essential operations of computer hardware such as simple memory and multi-bit variable manipulation.

This will provide an introduction to the way in which the complicated requirements of computer hardware are achieved by simple digital logic functions. The types of logic circuit that we will discuss here are broadly classified as *sequential logic*.

2.1   Flip-flops and latches

The basic requirement of a *memory* element is that it is able to store data presented to it, in a stable form suitable for subsequent recall. In a binary logic system, the basic unit of information or data is a single binary variable, hence the basic unit of memory is an element which can adopt and maintain one of two stable states (Fig. 2.1). The **NAND** operation can be regarded as a controllable *gate* (Fig. 2.2); when the **A** input is high, the output is the negation of the **B** input but when the **A** input is low, the output is logic **1**, regardless of the state of **B**. When the circuit of Fig. 2.1 is implemented with **NAND** gates, it can adopt two stable states while the extra inputs are at logic **1**, yet, by forcing one or the other of these low, it can be forced into one or other of its stable states. Since it is then possible to write to, or to read from, this simple bistable circuit, it represents a binary memory element.

The truth tables used in Chapter 1 can predict the operation of this circuit but do not provide a convenient depiction of its operation. Since the time history of the logic inputs is now important, it is more convenient to use a *timing diagram* (Fig. 2.3). The overall behaviour of the circuit can be generalised from the diagram.

> If both control inputs are left at logic **1**, the circuit rests in one of its
> stable states.
> If the $\bar{\text{S}}$ control input is taken to logic **0**, the **Q** output goes to logic
> **1**. If $\bar{\text{S}}$ then returns to **1**, the forced state of **Q** will persist.
> If the $\bar{\text{R}}$ control input is taken to logic **0**, the **Q** output goes to logic
> **0** and is left in that state when $\bar{\text{R}}$ returns to logic **1**.
> If both inputs go to logic **0** together, both outputs go to logic **1**. If
> both inputs return to logic **1** precisely together, the final state of
> the outputs will be indeterminate.

The overall effect is to *latch* the outputs into one or other state depending on which input last went to logic **0**. In the circuit shown, the outputs are latched by low levels on the control inputs. Since these are active when low, it is conventional to label or name them as negations, $\bar{\text{R}}$, $\bar{\text{S}}$.

Figure 2.1   A simple bistable circuit which can maintain two stable states:

$$\text{Q}=0, \quad \bar{\text{Q}}=1$$

or

$$\text{Q}=1, \quad \bar{\text{Q}}=0$$

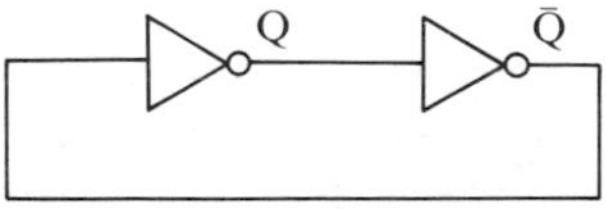

Figure 2.2   A binary memory element. The control signals are labelled with their conventional names, SET ($\bar{\text{S}}$) and RESET ($\bar{\text{R}}$).

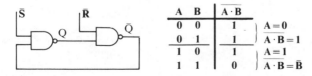

A	B	A·B	
0	0	1	$A=0$
0	1	1	$A \cdot B = 1$
1	0	1	$A=1$
1	1	0	$A \cdot B = \bar{B}$

In the early days of electromechanical relay logic circuits, this type of circuit was called a *flip-flop* due to the distinctive noise that it made.

The circuit described above is a basic *R–S flip-flop*, the inputs provide a SET or RESET operation of the outputs. In more recent times the name flip-flop has tended to be restricted to more elaborate logic circuits. The circuit of Fig. 2.2 is now usually called an *R–S latch*.

This simple circuit can be refined into a more convenient memory element by adding a few more gates as shown in Fig. 2.4. The three extra **NAND** gates allow the simple memory circuit to respond to a single input D (Data) and to latch the input at D only when the C (Clock) input is high. The word *clock* is used in computer terminology not only for a regularly 'ticking' timing source, but also for a logic pulse in general timing or latching operations. In the circuit here, the prime purpose of the timing input is to determine the instant or 'time' when data is to be latched. Thus this input is generally called the clock input. It is sometimes called by the more descriptive name of *Trigger* (T) input.

The timing diagram in Fig. 2.4 depicts how the clock, when low, prevents any change at the Q output. When the clock is high, the input is passed through to the output and 'follows' any changes there. When the clock input goes low again, the last input state is latched at the outputs. For this reason the circuit is sometimes called a *transparent D-latch*.

Set–Reset (S–R) inputs can be added to override the clocked inputs and force a known state. (Dashed on the figure.)

Figure 2.3   A timing diagram for the basic memory element. Time flows from left to right; the output states adopted in response to all possible changes in the input conditions are depicted. The initial condition is determined by the previous history; this indeterminacy is usually indicated as shown.

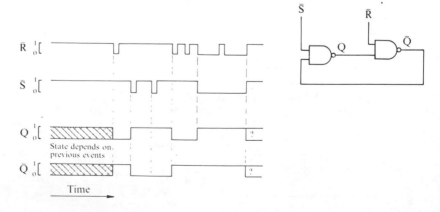

Simple latches or flip-flops based on the basic circuit of Fig. 2.4 suffer from a rather serious flaw when several of them are connected together. The circuit shown in Fig. 2.5 uses rectangular symbolic blocks (*black boxes*) to represent the logic diagram of Fig. 2.4. When the clock signal is raised to latch the data into the first flip-flop of Fig. 2.5(*a*), the output follows (almost) immediately and chains into the subsequent input so that the data input at the first flip-flop will scurry through the entire set to the last output. In the circuit of Fig. 2.5(*b*), a change at the output is fed immediately back to the input and the flip-flop will break into uncontrollable oscillation while the clock input is high. These are called *race conditions*. They may not have been what the designer desired of the circuits!

The race condition is avoided by the internal logic circuitry of two classes of flip-flop, the *master–slave flip-flop* and the *edge-triggered flip-flop*.

The master–slave circuit is shown in Fig. 2.6. It can be made from two simple R–S flip-flops with the second controlled by the negation of the first's clock input. While the clock input is high, the data at the input is passed continuously into the master flip-flop, but the slave flip-flop does not respond. When the clock input goes low, the master flip-flop is isolated from its input but stores the data present at that time. The inverted clock signal allows the slave to accept the master's output and transfer that data to its output but prevents the input data from racing through to subsequent flip-flops operating from the same clock signal. The overall effect is that the input data is latched over to the output at the instant when the master clock

Fig. 2.4   A 'clocked' data latch.

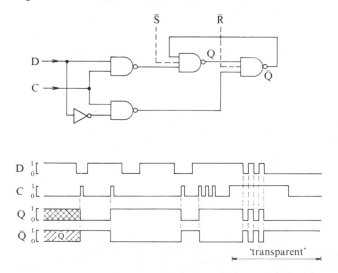

pulse goes low. The data at the master input must be stable immediately before (and during) the sensitive, negative-going clock edge.

The edge-triggered flip-flop circuit uses a subtle technique involving the propagation delay within its internal gates. A typical circuit for an edge-triggered D flip-flop is shown in Fig. 2.7. Gates G_4 and G_5 are a simple bistable output latch. The other four gates use the input clock pulse (C) to generate a single, very short pulse to one of the inputs to G_4 or G_5; this *trigger* pulse forces the output latch into its required state. A race condition is prevented by the propagation delays of the trigger pulse and of the output bistable latch. The input data D need only be stable for a brief interval before and during the transition. The internal operation of the circuit is left as an exercise for those who like convoluted logic.

MIL-STD-806 defines schematic symbols for flip-flops (Fig. 2.7). A rectangular box is used to represent many different compound logic

Figure 2.5   Circuits which demonstrate race conditions in simple flip-flops.

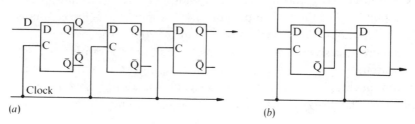

(*a*)   (*b*)

Figure 2.6   A simple master–slave D flip-flop.

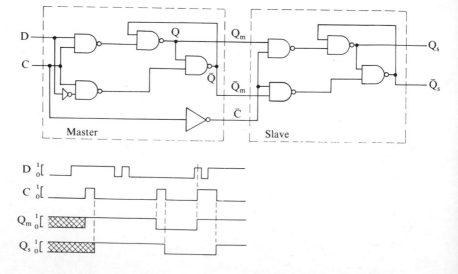

operations but the placement of inputs and outputs is defined for each type of flip-flop. Note the triangle symbol at the C input. This signifies an edge-triggered input. As shown, this input responds to a $0 \rightarrow 1$ transition (positive-going edge). An edge of opposite polarity has no effect on the outputs. A negative circle outside the box at the clock input indicates a sensitivity to a $1 \rightarrow 0$ transition (negative-going edge).

The **J–K edge-triggered flip-flop** is a general-purpose and very useful circuit that can perform many different operations. It usually has five inputs, as shown in the schematic symbol in Fig. 2.8.

The overall operation of the J–K flip-flop can be fully represented by timing diagrams but its response is commonly represented on a modified form of truth table as shown in Fig. 2.9. Table (*b*) depicts the response of

Figure 2.7   A positive edge-triggered D flip-flop made from **NAND** gates. Its standard schematic symbol is shown on the right.

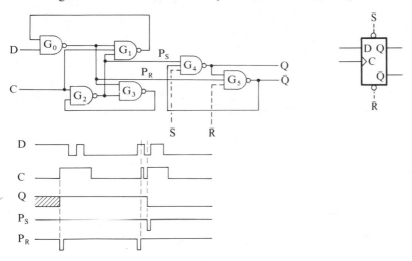

Figure 2.8   The symbolic diagram of a positive edge-triggered *J–K* flip-flop.

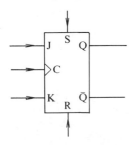

asynchronous active high S–R inputs. The state of the J–K inputs has no effect (**X** means either **0** or **1**), and the R–S inputs force and hold the outputs. The last row reflects the indeterminant condition found with the first flip-flop that we investigated.

Table (*a*) is coded in a way which clearly indicates the output response to different input conditions at the J–K inputs when both of the R–S inputs are inactive and the clock edge occurs.

The first output column (Q_n) indicates the output state before the clock edge, the second output column (Q_{n+1}) indicates the output state after the edge.

The first row indicates that the outputs are unaffected by the clock pulse when both J and K are low.

The second and third row indicate that, when the J and K inputs are complementary, the data input is latched over to the output by the rising clock edge.

The fourth row indicates that when the J and K inputs are both high, the two outputs *toggle*, that is, they *change state* at the clock edge regardless of their previous state.

A wide variety of flip-flop types are available in most logic families; most of them can be synthesised from the J–K flip-flop. We will not discuss them in detail; the discussion above should provide enough information to make all of them understandable from their manufacturer's data manuals.

Figure 2.9   The truth table for a *J–K* flip-flop with *high-active* R–S inputs; * indicates instability – does not persist when **R** or **S** go to **0**.

inputs				Before C↑		After C↑	
S	**R**	**J**	**K**	Q_n	$\overline{Q_n}$	Q_{n+1}	$\overline{Q_{n+1}}$
0	0	0	0	Q_n	$\overline{Q_n}$	Q_n	$\overline{Q_n}$
0	0	0	1	Q_n	$\overline{Q_n}$	0	1
0	0	1	0	Q_n	$\overline{Q_n}$	1	0
0	0	1	1	Q_n	$\overline{Q_n}$	Q_n	Q_n

(a)

J	**K**	**S**	**R**	Q_n	$\overline{Q_n}$
X	X	0	0	Q_n	$\overline{Q_n}$
X	X	0	1	0	1
X	X	1	0	1	0
X	X	1	1	1	1

(*b*)

There is no strict convention for the names of all flip-flop inputs. The S–R inputs are sometimes called Preset–Clear and the Clock input is sometimes called the Trigger input. There is no ambiguity, however, in the J, K and D input names.

Box 2.1
The flip-flop family

Many sequential logic operations can be performed by the J–K flip-flop. Other flip-flop operations can be generated by skilful use of its five inputs. There are, however, several specialised circuits commonly available in most logic families. A brief selection of these from the LSTTL logic family is presented in this box.

The humble S–R latch (basic flip-flop) is available in a package of four as the 74LS279. Fig. 2.10 shows the internal logic schematic diagram and the *pin-out* of the *chip*.

The 74LS74 is a dual, positive-edge triggered D flip-flop with active-low set and reset override inputs. Fig. 2.11 shows the symbolic diagram and truth table in the form presented in data manuals.

A wide variety of single and multiple J–K flip-flops are available with various input configurations. There are positive- or negative-edge triggered types and there are arrays with shared clock and/or R–S inputs. The 74LS112 is a dual negative-edge triggered J–K with fully independent inputs while the 74LS109 is a dual positive-edge triggered flip-flop with

Figure 2.10　The 74LS279 quad S–R latch.

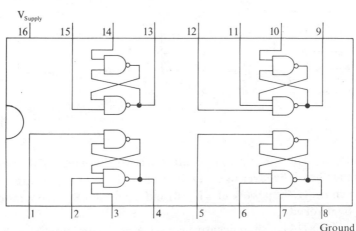

Figure 2.11 The 74LS74 dual D flip-flop: * indicates an unstable state which will not persist when $\bar{S}$, $\bar{R}$ go high; † indicates no change.

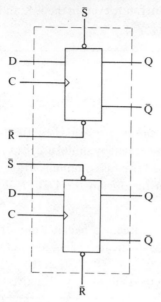

Inputs				Outputs	
$\bar{S}$	$\bar{R}$	C	D	Q_{n+1}	$\overline{Q_{n+1}}$
0	1	X	X	1	0
1	0	X	X	0	1
0	0	X	X	1*	1*
1	1	↑	1	1	0
1	1	↑	0	0	1
1	1	↓	X	Q_n†	$\overline{Q_n}$†

Figure 2.12 Some J–K flip-flop configurations.

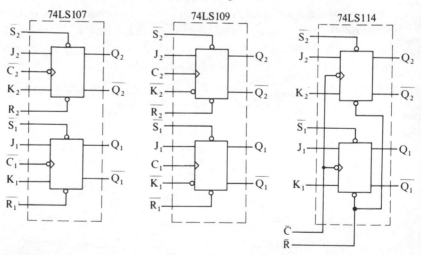

J–$\bar{K}$ inputs. This J–$\bar{K}$ input configuration is worthy of note; if the inputs are tied together, it behaves like a D flip-flop while, if the inputs are operated independently, the passive ($J = K = 0$) and toggle ($J = K = 1$) operations can still be invoked.

The 74LS114 is a dual negative-edge J–K with common C and R inputs but independent S inputs. The 74LS107 is a dual J–K master–slave flip-flop.

2.2 Multi-bit data manipulation in registers

This section explores the application of flip-flops to several important data operations required within a computer.

2.2.1 Multi-bit data register

A multi-bit *data latch* or *data register* is used to store the data associated with a multi-bit binary variable. It can be made from an array of J–K or D flip-flops. The circuit of Fig. 2.13 shows four J–K flip-flops arranged as a 4-bit edge triggered data latch. It is able to latch and store the data pattern present at its inputs at the instant when the common clock line goes high. It will be shown in Chapter 4 that this function is essential for the operation of a computer; data passed between the elements of a computer are only provided by the source for a very short time interval. The receiver of the data must be able to 'pick-off' or *latch* the data during this brief interval.

Box 2.2
Data sources and destinations

This box describes a simple system which allows a Master unit to send multi-bit data words to several different destinations.

Fig. 2.14 shows a block schematic diagram of a system where the three black boxes labelled L_A, L_B and L_C each contain the multi-bit latch of Fig.

Figure 2.13   A 4-bit edge-triggered latch.

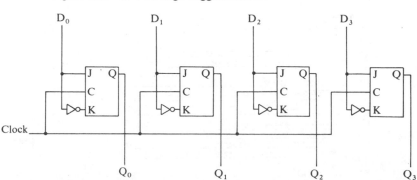

2.13. The other black box on the left, the Master block, is a more complicated circuit which can generate 4-bit data patterns on the data lines and can generate three separate clock pulses.

Integrated circuits at the MSI (Medium Scale Integration) level are available for multi-bit data latching. In the LS family the 74LS75 is a 4-bit transparent D latch. The 74LS175 and 74LS174 are respectively 4-bit and 8-bit positive-edge triggered latches (flip-flops).

Most of the data transactions in a computer are conducted over special communication lines which require special logic output characteristics. This will be described in detail in Chapter 4; the 74LS373 and 74LS 374 8-bit data latches provide these special characteristics and will be described there.

Figure 2.14   A data distribution system. The timing diagram shows a sequence of signals generated by the Master block which causes three different data patterns to be stored at each of the latches. Note that, although only one 4-bit word can be sent at a time, the master is able to manipulate the data in all three latches by *time-multiplexing* data transactions over the single set of data lines.

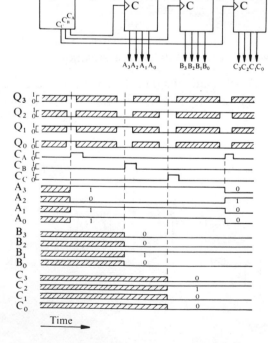

2.2.2   Multi-bit shift registers

As its name suggests, a *shift register* is a register or latch which provides the capability to shift its multi-bit contents to the left or to the right. Master–slave and edge-triggered flip-flops (which avoid the race condition) can be inter-connected to carry out these operations.

Fig. 2.15 depicts a 4-bit right-shift register. At each clock edge, the data present at each D input is latched over to its respective output; the time delay across the edge-triggered flip-flop ensures that the data appears at each output *after* the following gate has latched its input. This ensures that the data pattern is moved just one place right at each rising clock edge.

This operation is of considerable importance in logic and arithmetic manipulations of multi-bit variables within a computer. It also allows data presented in a serial stream at the D input to be converted into a parallel multi-bit word at the Q outputs. (We discuss this further in Section 2.4 and in Chapter 8.)

Further logic gates added to the circuit of Fig. 2.15 allow multi-bit data to be latched into the register and then to be shifted right. Fig. 2.16 shows a suitable circuit. Note that a *serial* output taken from Q_3 represents the

Figure 2.15   The schematic diagram of a 4-bit shift register made from edge-triggered D flip-flops; the data moves one place to the right (on the diagram) when the clock edge rises.

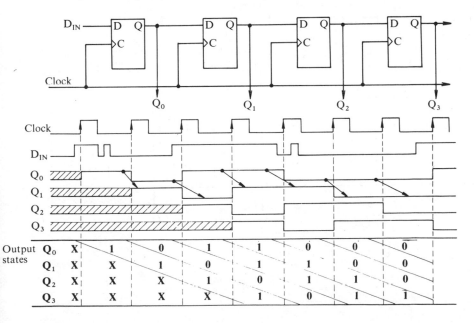

complement of the *serial to parallel conversion* discussed before. This provides *parallel to serial conversion* of data for computer communications.

The circuit can be further enhanced to allow data to be shifted right or left under the control of another input signal; this is left as an exercise.

Box 2.3
A brace of shift registers

A wide variety of shift registers is available in most logic families. A selection of those available in LSTTL is described in this box.

Figure 2.16   A latch–shift register. Note that the control signal LATCH modifies the response to the clock signal. When high, it presents the multi-bit input data to the flip-flops; when the clock input goes high, the data present will be latched in. When the LATCH is low, the register performs a right-shift of its data on the rising clock edge.

The truth table depicts the circuit's operation in a very compact form. It is sometimes called a *transition table* and is commonly used in data manuals. A wide variety of tables and diagrams are used in data manuals; they are invariably self-explanatory.

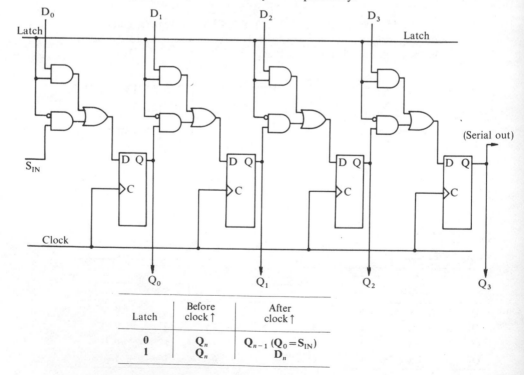

Latch	Before clock ↑	After clock ↑
0	Q_n	Q_{n-1} ($Q_0 = S_{IN}$)
1	Q_n	D_n

The 74LS164 is a simple 8-bit shift register which operates very like the one shown in Fig. 2.15. It has a single asynchronous active low clear (or reset) input which overrides all other inputs and forces all outputs to the low state. Data must be loaded into the register in serial manner, but all eight outputs are available in parallel.

The 74LS166 is also an 8-bit simple shift register but its inputs are arranged for serial output. Data can be latched into it either serially or in parallel through individual inputs, but the stored data can only be output through the single serial output.

The 74LS195 is a straightforward 4-bit shift register with parallel input and parallel output as depicted in Fig. 2.16. It also has serial data inputs in the J–$\bar{\text{K}}$ configuration discussed for the 74LS109 flip-flop in Box 2.1. This register is thus very versatile and can be used for data latching, serial to parallel conversion, parallel to serial conversion or simple right shifting. It also has a single asynchronous (overriding) CLEAR (reset) input.

The 74LS194 is a fully accessible bidirectional 4-bit shift register. It is like the 74LS195 but can move data in either direction using separate left and right serial D inputs instead of the J–$\bar{\text{K}}$ pair.

The complex logic operations represented by these registers do not yet have universally accepted standard schematic symbols. Their internal logic can always be depicted using diagrams composed of standard symbols but these are too complex for routine use. A rather loose convention has evolved to represent them by a simple rectangular black box marked with the family code number, and to indicate the input and output connections by using their standardised symbols. A typical representation is shown in Fig. 2.17. Manufacturers tend to use different formalisms in their data

Figure 2.17   A typical symbolic diagram for a shift register.

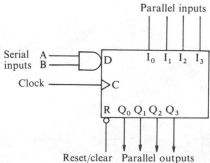

manuals for representing circuit operation via truth tables etc. They are, however, always easy to understand.

2.3   Sequential control circuits

A sequential circuit, like a combinatorial circuit, has multiple inputs and outputs; the output state, however, depends on the *time-history* of the inputs. This section looks at some very simple examples. A formal study of the theory and of design procedures is not relevant to this book; references are given for further reading at the end of this chapter.

Fig. 2.18 depicts three cascaded J–K flip-flops connected for toggling. Whenever the clock input goes positive, the Q_0 output changes state; with a repetitive clock signal, it represents a signal at half the clock frequency. The two other outputs, Q_1 and Q_2, represent further divided clock signals. As time goes by, all eight possible states of the three outputs will be generated in a well-defined and repetitive manner. The exploitation of this circuit as a counter is explored in Exercise 2.5.

The circuit of Fig. 2.18 has a rather serious flaw. The propagation delay across the flip-flops delays the response of the output; with cascaded output/input connections, the change of state in response to a clock input can take some time to *ripple through* the circuit. It is called an *asynchronous counter*. During the ripple propagation time, the outputs are scrambled. The flaw can be eliminated by a more elaborate circuit called a synchronous counter. The 74LS161 chip (Fig. 2.19) is an example; analysis of its operation is left as an exercise.

Well-defined control signal sequences are needed in many systems. In Chapter 5 we will be exploring the internal operation of a microprocessor; the timing diagram depicted in Fig. 2.20(*a*) will be used there. We will end this section by presenting a simple sequential circuit based on the 74LS161 mentioned above, to generate the required signal sequence. Analysis is left as an exercise.

Figure 2.18   An eight-state sequential circuit made from three flip-flops.

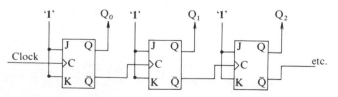

2.4   Linear sequential circuits and digital communications

Communication systems generally need a large vocabulary of predefined symbols. A glance at a typewriter keyboard shows that common, written language communication needs about 80 symbols (including numerals and upper/lower case letters).

Figure 2.19   The internal logic of the 74LS161 4-bit synchronous counter (taken from the National Semiconductor Logic Data Manual).

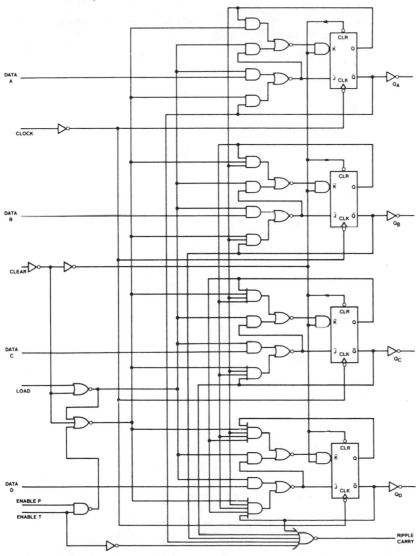

For communication via a binary system, the symbol vocabulary must be *encoded* into multi-bit binary (digital) words. In the case of written language, seven bits are essential (see Box 2.4). The general problem in digital communication is to transmit and receive information through a medium in the form of multi-bit digital words. Media such as the telegraph system described briefly in Chapter 1 are only able to be in one of two states at any instant, hence the multi-bit binary words must be transmitted and received in a *bit-serial* manner. A brief mention of this was made in Section

Figure 2.20 (*a*) A set of seven repetitive pulse signals and (*b*) a simple circuit to generate them.

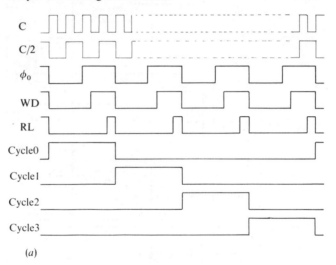

(*a*)

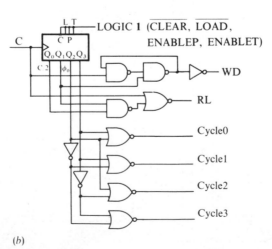

(*b*)

2.2.1, where registers which could convert a multi-bit word into a serial string for the transmitter and vice versa for the receiver were described.

If the medium is likely to add *noise* to the data signal, there is a possibility of digital errors occurring; a change of state of even a single bit of a transmitted word will result in a false symbol being received. In many critical applications, the probability of such errors must be reduced to a very low level. Where signal transmission errors are unavoidable (for example, in long-distance radio links etc.) the only alternative is to develop logic systems which can at the very least detect and, if possible, also correct the random, induced errors. This is achieved by introducing additional redundant information into the symbol coding; the extra information can be used to verify valid codes and to correct errors.

If a lot of redundancy is included, it is easy to detect and correct a faulty pattern to any predefined level of reliability. A simple example of this is to repeat every character three (or more) times and to make the receiver *poll* the received code. This simplistic solution considerably reduces the information rate. A trade-off must be made between error detection/correction performance and speed. The optimum performance can only be achieved if the redundant coding is made as efficient as possible. The work of R. Hamming in 1950 led to the development of the efficient *Hamming code* schemes.

The need for efficient *error detection and correction logic* systems led to a considerable body of development work with sequential flip-flop arrays. The problem is handled in a manner very similar to that discussed briefly in Chapter 1 for minimisation of combinatorial logic.

A considerable body of elegant theoretical work has been carried out in the field of *linear sequential logic* circuits using flip-flops. The field's name derives from the way in which sequential binary patterns are represented as simultaneous linear polynomial equations and are manipulated and analysed directly by matrix methods. This allows minimal expressions which in turn achieve optimal error handling with minimum redundancy. The optimised systems could then be implemented and generated economically by sequential circuits using shift registers. Since the development of microprocessors, the problems of encoding, decoding, error detection and correction and, recently, data encryption have been transferred almost completely to computer software. The development of compact and cheap integrated circuits has made the minimisation of sequential logic circuits relatively unimportant.

Within the interior of a microcomputer, data paths are kept short and noise-induced errors are avoided by good design of the data pathways (this will be discussed in Chapter 4). In very large memory arrays, however, even a very small bit-error probability becomes significant; error detection and correction techniques are becoming common (see Chapter 7).

We need go no further into linear sequential theory to understand fully a computer's operation; introductory references for further reading are given at the end of this chapter.

Box 2.4
Computer coding of common language text

The 52 alphabetic, 10 numeric and 15 or so punctuation or format symbols of common written text require a binary code with at least seven bits. An 8-bit word was introduced by IBM as a convenient size for *Binary Coded Decimal* (*BCD*) numbers (see Chapter 5). IBM use their own code called *Extended Binary Coded Decimal Interchange Code* (*EBCDIC*) for text symbols. It uses all eight bits but only defines about 170 of the 256 possible patterns for text symbols, a wide range of special symbols and *non-printing* control codes. The control codes are used by the communication control logic to monitor message beginnings, ends etc.

Most mini- and microcomputers use another code defined by the American Standards Committee for Information Interchange (ASCII). It is a 7-bit code and defines all 128 possible patterns for text and control codes. The use of the eighth bit is not defined in this standard.

The International Standards Organisation has defined a further code called *International Alphabet 5* (*IA5*); this code has been adopted by the *CCITT* (*International Consultative Committee on Telegraphy and Telephony*). IA5 is identical to ASCII except for some allowance for local variation of currency symbols etc.

Microcomputers commonly use ASCII code (or a subset of it) for input of text from their keyboards etc. and for internal text storage. The control codes are usually redefined for internal computer control functions.

When the 7-bit ASCII is filled out to an 8-bit word the redundant bit can be set high or low, or can be used to add some redundant data to the code for error detection (as mentioned before). The eighth bit usually holds the sum (modulo-2) of the other seven bits. This redundant bit is called the *parity bit*. The 7-bit code

1011110

thus would be augmented to

11011110

and called *even parity* since the sum of all eight bits is an even number (**0**); *odd parity* is also used.

When this augmented, 8-bit code is transmitted, a receiver system can check the parity of the received word and can then detect if an odd number of bit errors have occurred (it cannot detect an even number of errors). The statistical distribution of multi-bit errors makes this scheme useful only for detecting single-bit errors where the overall error probability is low (that is, where multi-bit errors are very unlikely). Codes with more redundancy are needed for serious error detection and correction of noisy signals.

Microcomputers usually ignore the parity bit or use it for internal computer functions.

2.5   Mass memory circuits

This chapter was introduced with a simple memory element based on **NAND** gates. This was to demonstrate the concept of a memory cell within the binary logic operations defined in Chapter 1. A major limitation of a computer is the amount of memory that it contains, thus it is essential to provide the largest practicable array of memory elements; this *mass memory* cannot be provided economically by enormous numbers of **NAND** gates. Many ingenious techniques have been devised and developed for the computer's mass memory. The continuous pressure to provide faster and bigger memory arrays has kept the computer industry at the very limit of available technology since the mid-50s. There is neither the space nor the justification here to include an historical summary of this. This short section will only describe a couple of techniques used for the provision of *internal, directly accessed* memory within a contemporary microcomputer and the treatment here will be limited to a brief description of the actual memory cells used. The techniques used to write and read information to and from these cells will be left until Chapters 4 and 8, where the data pathways and memory organisation within the microcomputer are described.

Economical mass storage requires a simple and economical basic memory element. A simple bistable memory element analogous to that of Fig. 2.1 can be made from a pair of bipolar or field effect transistors. The circuits in Fig. 2.21 show the basic minimum configuration.

Fig. 2.21(*a*) shows a bipolar transistor cell. The current gain of the transistors ensures that only one will be in saturation and the other will be turned off. Both alternative states are stable and will persist indefinitely while the supply voltage is maintained. This circuit is the basis of all bipolar

Figure 2.21   Basic circuits for (*a*) bipolar and (*b*) MOS field effect memory cells.

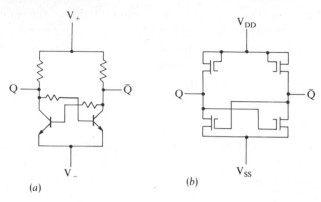

(*a*)   (*b*)

static memory integrated circuits. TTL technology allows relatively simple reading and writing logic techniques with very fast response (of the order of 20 ns). The circuit of Fig. 2.21(*b*) shows the same basic cell with Metal Oxide Silicon (MOS) field effect transistors. The most obvious difference is the elimination of the current limiting resistors. The need for only one type of device in the cell greatly simplifies the production process and the very compact layout of MOS field effect transistors allows very dense placement of these simpler cells. The trade-off is speed; MOS devices made with simple cheap technology cannot achieve much better than about 150 ns response time.

The very low gate leakage current of MOS transistors is used to make very high density memory chips. The binary information is not stored in a true bistable element, but as a high or low charge in the transistor gate capacitance. This allows a single bit of information to be held in a single transistor and allows over 256 000 bits of data to be stored in a single integrated circuit chip of about 5 mm square (the bit cells are less than 10 microns square). The million-bit chip is becoming available at the time of this book's publication. The cell is not a bistable element; the tiny data charge leaks away in a few milliseconds. Each and every cell must be read and rewritten (refreshed) before its charge decays. This is called dynamic memory.

The organisation and operation of various types of mass memory such as *Read-Only Memory* (*ROM*) and *Programmable* and *Erasable Read-Only Memory* (*PROM* and *EPROM*), will be described in Chapter 8.

Exercises

2.1 Explore the behaviour of a primitive memory cell as in Fig. 2.2, but made from **NOR** gates.

2.2 If you have access to a digital logic demonstration set, wire up and test the circuits of Figs. 2.2, 2.4, 2.6 and 2.7.

2.3 Analyse the behaviour of the circuit below with respect to propagation delays, when the C input changes from logic **0** to **1**.

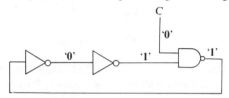

2.4 Analyse the circuit of Fig. 2.7 to determine the effects due to propagation delay in the logic gates.

2.5 (*a*) Review the toggling action of the JK flip-flop when both the J and K inputs are held at logic **1**. Note that it can be considered as *counting* (modulo-2). Analyse the circuit of Fig. 2.18 to confirm that the second flip-flop will count modulo-4 and so on. Generalise this scheme to understand the purpose of the full circuit. It is called a *ripple-carry counter*.
(*b*) What would be the effect of changing each clock input to the $\bar{Q}$ outputs of its preceding flip-flop?
(*c*) Draw a timing diagram showing the three outputs of the ripple-carry counter (Fig. 2.18) with a 10 MHz input clock and a 20 nS gate delay. Tabulate the three outputs to show the 'glitches'.
(*d*) Study the internal circuitry of the 74LS161 4-bit counter, to see how the counting glitches are eliminated in this *synchronous counter*. Look into the manner in which these counters can be cascaded. (Refer to the data manual.)
(*e*) Confirm that the circuit of Fig. 2.20(*b*) generates the pulse sequences of Fig. 2.20(*a*).

2.6 If you have access to a suitable digital logic demonstrator, wire up and test the circuits of Figs. 2.13, 2.15 and 2.16.

2.7 Design one stage of a logic circuit which when implemented on all stages of the circuit of Fig. 2.16 allows the shift register contents to be moved left or right dependent on the state of a further control signal.

2.8 Show that the circuit shown below tests for even parity of the three inputs. Extend this circuit to test for even parity of four bits and then for five bits.
   Can the circuit be used to *generate* an even parity bit?

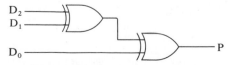

References and further reading

1 *High Speed Computing* (ibid. Chapter 1)
2 *Computer Architecture and Organisation* (ibid. Chapter 1)
   References 1 and 2 describe some early memory techniques.
3 'A Method for Synthesizing Sequential Circuits', G. H. Mealy, *Bell System Technical Journal*, **34**, 1955, pp. 1045–79
4 *Introduction to Switching Theory and Logical Design* (ibid. Chapter 1)
5 *Linear Sequential Circuits*, A. Gill, McGraw-Hill, New York (1967)
   References 3, 4 and 5 cover the development of sequential circuit analysis.
6 *LSTTL Data Manuals* (Fairchild, Motorola, National etc.)

Up-to-date reference and data manuals for LSTTL, ALS, CMOS etc. integrated circuits show the very wide range of available integrated circuits (Ref. 6).

3

Numbers and arithmetic in binary logic

Preamble

Chapters 1 and 2 have demonstrated how binary logic operations are implemented in digital electronics. Computers are nothing more than rather complicated machines which carry out a range of such operations on a large array of binary logic variables. In many applications they must be able to perform mathematical operations; this requires that these be formulated in terms of binary logic operations.

We have seen how processes expressible in terms of logic, can be implemented in digital electronic hardware. This chapter will show how binary logic can be used to *model* numbers and the fundamental arithmetic operations. The significance of the model to the design of the computer's *architecture*, and the manner in which the model is implemented, will be described in later chapters.

Arithmetic operations performed by software often impose the principal restraint on the speed of microcomputer control and data collection systems. The treatment in this chapter is deeper therefore than is needed for a basic understanding of computer software operation. The extra information is intended as the starting point for the design of fast, dedicated, arithmetic, peripheral hardware; some simple applications will be described in Chapter 9.

3.1   Numbers as binary variables

A single logic variable can adopt only two *states*; it is quite unsuitable to model a variable which can adopt many states. An array of n individual variables, however, can adopt 2^n different states. (See Exercise 1.6(*b*).) A variable system with any finite number of states can therefore be modelled by using a suitable array of binary variables. Contemporary microcomputers use an array of either eight or sixteen binary variable *bits*. The number of bits in the array which the computer handles is commonly called the *word-size* and the array of individual bits is called a *word*. With an *8-bit word*, 256 states are possible; with a *16-bit word*, there are 65 536 states.

A finite set of numbers can be modelled by a multi-bit binary word if each state is used to represent one number of the set. Some form of assignment or coding is needed; the common *place-value notation* used for decimal numbers is very well suited to binary variables. The conventional decimal numbering convention assigns a specific value to the position of each variable in the number (see Box 2.1). It uses digits (from Latin, *digitus* – finger or toe) with ten states symbolised by the numerals 0–9. This is one of many examples of confusing nomenclature in computer jargon. In Chapter 1 the word *'digital'* described *binary* logic systems. From here on it will be up to you to decide whether to respond to 'digit' by contemplating ten digits or just two.

The concept of place-value notation for numbers seems to have developed slowly during the Third Millenium BC in Babylonia, to simplify recording both large and small financial transactions. The currency, and therefore the number system, was sexagesimal (base-60, easily divisible by 2, 3, 4, 5, 6, 10, 12 etc!). The Babylonians used multiplication tables tabulated up to 60 times 60. During the Second Millenium BC they introduced a symbol for zero. The motivation seems to have been the elimination of the ghastly accounts errors that would result from dropping or gaining a digit with a base-60 place-value number.

The importance of the new symbol in arithmetic computation was gradually realised and the improved number system became the standard for Byzantine, Ancient Greek and Islamic arithmetic. Islamic scholars introduced a decimal place-value number system but used both sexagesimal and decimal numbers. From the Fifth Century AD, Hindu astronomers adopted the alternative Islamic decimal number system; their motivation can be assumed to be the relative ease of memorising the multiplication table. The decimal system came to Europe from the Hindus by way of Arabia. Our present numerals evolved during the First Millenium AD from the Arabic signs for the ten digits [*sic*].

But we digress! The place-value notation for number systems can easily be adapted to two-state (binary) variables. This chapter will explore the simplifications that result by changing from a decimal to a binary number system. It will be obvious that the motivation for the change is in the relative ease with which arithmetic operations can be expressed as simple *logic* operations on binary variables. The power and efficiency of a binary logic basis for computer architecture will become obvious. From here on, *where it is not obvious*, base-10 numbers will be designated by a subscript, viz.

$$12345_{10}$$

Box 3.1
Binary and decimal numbers

Consider the ordered array of decimal digits below which represent a decimal number in common place-value notation.

12345.678

The meaning of this array of digits is actually

$$1 \times 10000 + 2 \times 1000 + 3 \times 100 + 4 \times 10 + 5 \times 1$$
$$+ 6 \times 1/10 + 7 \times 1/100 + 8 \times 1/1000$$

(Please excuse the use of place-value notation to describe place-value notation; we must avoid a further digression, this time into meta-languages.)

In decimal exponent notation this is

$$1 \times 10^4 + 2 \times 10^3 + 3 \times 10^2 + 4 \times 10^1 + 5 \times 10^0$$
$$+ 6 \times 10^{-1} + 7 \times 10^{-2} + 8 \times 10^{-3}$$

Consider now the array of binary digits below. We will declare that it represents a binary number in place-value notation.

10110.011

By extrapolation from the decimal example, the meaning of this array is

$$1 \times 16 + 0 \times 8 + 1 \times 4 + 1 \times 2 + 0 \times 1 + 0 \times 1/2 + 1 \times 1/4 + 1 \times 1/8$$

In base-2 exponent form this becomes

$$\mathbf{1 \times 2^4 + 0 \times 2^3 + 1 \times 2^2 + 1 \times 2^1 + 0 \times 2^0 + 0 \times 2^{-1} + 1 \times 2^{-2} + 1 \times 2^{-3}}$$

The common place-value notation is obviously quite suitable and very convenient for representing numbers in terms of binary variables. It is only necessary to reduce the number of digits to two (**0** and **1**) and to assign place values in powers of the number base 2.

The concept of a decimal point, which allows fractional and irrational numbers to be properly represented, transfers neatly into the binary system in a form which could be called a *binary point*.

3.2   Addition with binary numbers

When confronted with the expression

$$853_{10} + 653_{10}$$

we have been taught to rearrange the arguments of the expression to implement our long-addition *algorithm*. The two numbers are first aligned so that equivalent place-value digits correspond and we then carry out a sequence of 'simple addition' operations starting from the least valued place. We use a large *look-up table* held in our memories as a two-dimensional array with entries for every possible pair of single digit additions; a *carry* flag or digit is used to note where a sum *overflows* into the next more significant place-value digit.

The large look-up table is rather awkward to implement in simple electronic digital logic. If, however, the same algorithm is performed on binary numbers, the *binary addition* look-up table becomes very simple (Fig. 3.1) and is easily implemented in simple boolean operations. The inclusion of the carry bit from the previous addition is easily expressed by a three-input truth table (Fig. 3.1(*d*)). Box 3.2 explores how both addition operations are implemented with digital logic gates.

Figure 3.1   (*a*) The binary long-addition algorithm, (*b*) the look-up table for two digits, (*c*) the look-up table expressed as a truth table, and (*d*) the truth table with a *carry* from the lower digit.

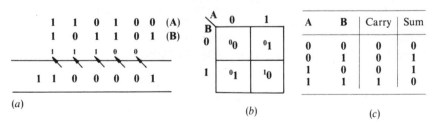

A	B	Carry	Sum
0	0	0	0
0	1	0	1
1	0	0	1
1	1	1	0

(*a*)   (*b*)   (*c*)

A	B	Carry in	Carry out	Sum
0	0	0	0	0
0	1	0	0	1
1	0	0	0	1
1	1	0	1	0
0	0	1	0	1
0	1	1	1	0
1	0	1	1	0
1	1	1	1	1

(*d*)

Box 3.2
Adders in digital logic

In this box we explore the implementation of binary addition in digital logic. We start by considering the truth table of Fig. 3.1(c) for two bits of the same place-value. The table shows the carry (C) and the sum (S) to be simply the **AND** and the **EOR** operations respectively. The diagram of Fig. 3.2 thus represents addition for two bits of equal value; it is called a *half-adder*. It can only carry out the addition of the least significant bits (l.s.b.s) of two binary numbers; for all other bits the carry must be included. The truth table in Fig. 3.1(d) defines the operation required. We will now develop the logic circuit required for this *full adder*.

First consider the carry-out signal, C_{OUT}; it is to be true whenever two or more of the three inputs are true. This can be stated in boolean logic as

$$\mathbf{A} \cdot \mathbf{B} + \mathbf{A} \cdot \mathbf{C}_{IN} + \mathbf{B} \cdot \mathbf{C}_{IN}$$

and can be implemented as in Fig. 3.3.

Now for the S output of the truth table. When C_{IN} is false, it is, as for the half adder, the **EOR** operation; when true, it is the negation. In boolean algebra

$$(\mathbf{A} \oplus \mathbf{B}) \cdot \bar{\mathbf{C}}_{IN} + \overline{(\mathbf{A} \oplus \mathbf{B})} \cdot \mathbf{C}_{IN}$$

Figure 3.2 (a) The schematic diagram for a half-adder and (b) its representation as a black box.

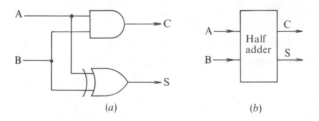

(a)   (b)

Figure 3.3   The carry-out for a full adder.

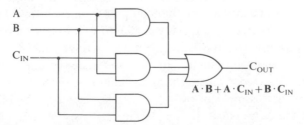

$$\mathbf{A} \cdot \mathbf{B} + \mathbf{A} \cdot \mathbf{C}_{IN} + \mathbf{B} \cdot \mathbf{C}_{IN}$$

Figure 3.4   The carry-out of a full adder using **EOR** gates.

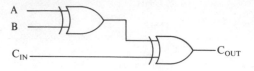

Figure 3.5   A full adder implemented entirely in **NAND** gates.

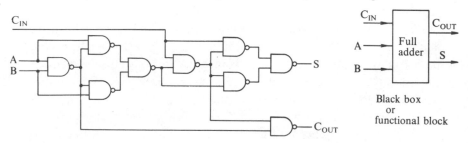

This can be reduced by inspection to the simple operation

$$\mathbf{C}_{OUT} = (\mathbf{A} \oplus \mathbf{B}) \oplus \mathbf{C}_{IN}$$

and can be implemented as in Fig. 3.4 using **EOR** gates.

Fig. 3.5 shows an economical configuration for a full adder made from **NAND** gates alone. The design strikes a compromise between a minimum number of gates and a short carry propagation path from input to output.

The terms *functional block* or *functional* are commonly used to describe a *black box* which performs a high-order logic task.

The full adder function is so simple to implement that the whole addition algorithm for two multi-bit words can be performed by a single functional block assembled from full adders (Fig. 3.6). Note, however, that any carry generated affects the outputs of subsequent stages. In the worst case, a carry resulting from the first stage can initiate a carry sequence which ripples through the entire chain. This is a similar effect to the simple counter discussed in Chapter 2. It is called a *ripple-carry* configuration. During the carry propagation time the S outputs from each stage may be wrong. To ensure a correct result, it is essential to wait for the worst case, ripple-carry propagation delay, before accepting the result; this time can be considerable for a multi-bit adder. The multi-bit adder can of course be considered and analysed as a multi-input functional block on its own account, not as an assembly of other simpler functionals. The boolean expression for the outputs can be represented by a set of simultaneous

Figure 3.6   A multi-bit adder assembled from full-adder functional blocks.

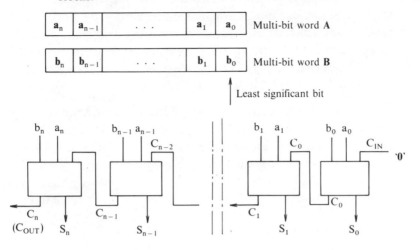

Figure 3.7   Symbols for a multi-bit adder. The one at the left is common in data books, the other is the more schematic form used in computer literature.

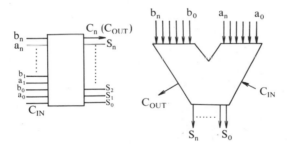

boolean equations. Manipulation of these provides explicit combinational equations for the C output and each S output.

Commercial gates are available in most logic families which minimise or eliminate the ripple delay in this way. These are generally called *look-ahead carry* adders. Two common symbols used for them are depicted in Fig. 3.7.

Box 3.3
Fast adders and look-ahead carries

Babbage (Section 1.2) recognised the serious propagation delay problem of the ripple-carry adder used in the Difference Engine and laboured long to overcome it in the design of the Analytic Engine. His solution involved performing the multi-digit decimal additions in two stages with a mechanism which he called the *anticipating carriage*. In the first stage of addition, the mechanism added the individual pairs of digits and signalled any actual or *anticipated* carries. (Any digit standing at '9' *anticipates* a carry out if a carry input is passed to it, and will propagate the carry.) In the second stage, every signalled carry could then be performed in a single step and propagated as a chained action.

In the LS family, the integrated circuit 74LS285 carries out a full addition between two 4-bit binary numbers with carry-input to generate all four sums and a carry-out. The schematic diagram of Fig. 3.8 depicts the logic. Note that the carry-out is generated through a sequence of three gates only and as a direct combinatorial operation from all nine input variables. Each of the four sum outputs are similarly derived from all lower value inputs.

Figure 3.8   A commercial 4-bit adder with 'look-ahead carry' (74LS285).

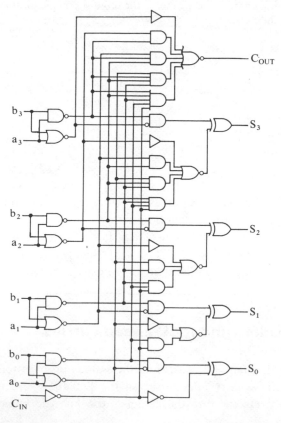

A full 8-bit or 16-bit adder based on this principle is included within the microprocessor of a microcomputer. We will see in Chapter 6 how it is included within the architecture of the machine.

3.3 Negative numbers and subtraction

A model for numbers is often described with a *number* line where each element is bigger than that to its left and smaller than that to its right. A negative number is coded by the special sign ' – ' in our ordinary decimal place-value number system; the number following this sign is the *magnitude* of the actual number. This creates a discontinuity at the origin; the addition algorithm must be modified to handle negative numbers. Subtraction can be defined by a new look-up table or it can be defined as the addition of signed numbers,

$$A - B = A + (-B)$$

Machines are not yet as smart as us; they cannot handle subtleties of meaning and complex symbolisms. Thus, when devising a machine to perform some tasks, it is essential to rationalise all of the processes involved; in moving from hand execution of decimal arithmetic to machine execution of binary arithmetic, the algorithms must be reconsidered in terms of machine constraints. The most efficient algorithms *for the machine* must be used. There are two major problems which must be solved for the machine.

(a) Decimal signed numbers need a new symbol, the ' – ' sign (we will consider the decimal-point symbol in Section 3.6). This is easily introduced in a conceptual system designed for people. However, in a binary logic machine, there is no simple way to introduce the new symbol; the concept of negative numbers must therefore be made implicit within the existing number representation.

(b) Our common decimal number system of arithmetic has no constraint on the magnitude of the numbers which can be defined; place-values can, in principle, be extended indefinitely in both directions. In machine implementation, however, the hardware limits the number of place-values available.

We will first examine how a simple decimal machine copes with these problems; we will then apply what we find to binary machines.

Consider again the motor car odometer used in Chapter 1. It is restricted to five (sometimes six) digits and can therefore only adopt 100 000 states, the values 00000–99999. It is, in fact, a *modulo-100 000* counter; when it exceeds its maximum value of 99999, it overflows its *number line* and registers 00000. If this odometer is moved backwards (subtraction), then when it goes below 00000 it registers 99999. This shows the natural way in which a machine wants to represent negative numbers. Fig. 3.9 shows the number line wrapped around upon itself. This number representation makes subtraction and addition equivalent. Consider the inferred subtraction of two numbers,

$2 + (-2)$

For the odometer number system, these two numbers become

$00002 + 99998$

and the answer, using the common addition algorithm, is

$(1)00000$

The addition/subtraction problem has been changed to one of deciding where positive numbers end and negative numbers begin on the cyclic number line of Fig. 3.9. The boundary was set arbitrarily at half-way. Note that the transition from 00000 to 99999 and vice versa is no longer an *overflow* or *underflow* of the number line, it is now a *change-of-sign*; *overflows* or *underflows* have been transferred to the arbitrary boundary where the greatest magnitude negative number changes to the largest positive number and vice versa.

Now we are ready to make the final jump to signed binary numbers. For simplicity in the diagrams used, we will use a 4-bit binary number line; there are no problems at all in extending it to any number of bits.

The sixteen states of the 4-bits can model a simple unsigned integer number

Figure 3.9   The number line for a decimal machine (odometer).

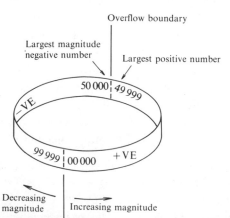

line as shown in Fig. 3.10(*a*). When we change to the signed model of Fig. 3.9, the number line of Fig. 3.10(*b*) results. The *most significant bit* (*m.s.b.*) of the place-value binary number now indicates the sign of the number! Thus we have not only a unification of addition and subtraction, but also a very convenient and explicit method for indicating the sign of the number. The number line overflow or underflow condition is again transferred from the origin to halfway around the ring (number line). This representation is called *two's complement*. A positive number is transposed into a negative number of the same magnitude by the simple algorithm of *negating* each of its bits and then adding one to the result. Box 3.4 demonstrates this and shows how subtraction can be performed by a multi-bit adder.

Figure 3.10　A 4-bit binary place-value number line; (*a*) unsigned, (*b*) signed and (*c*) the signed number ring.

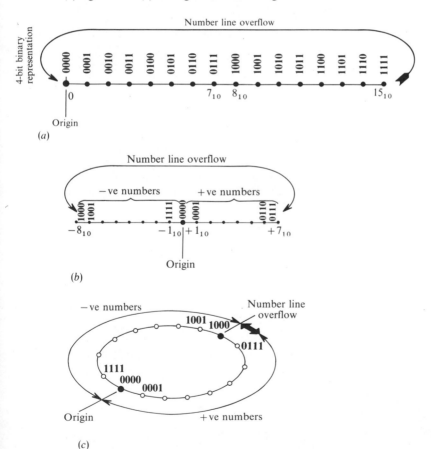

Box 3.4
Subtraction using digital logic

Example 3.1

Use the two's complement algorithm for negative numbers to add the numbers 5_{10} and -5_{10}. Show that the result is 0_{10}.

$$5_{10} = \textbf{0101}$$

We derive the two's complement of 5_{10} (i.e. -5_{10}) by negating its binary representation and adding '1'. (We will use the symbol $+_{ar}$ to distinguish addition from the logic operation **OR**.)

$$-5_{10} = \overline{\textbf{0101}} +_{ar} \textbf{0001}$$
$$= \textbf{1010} +_{ar} \textbf{0001}$$
$$= \textbf{1011}$$

The addition (the subtraction of 5_{10} from 5_{10}) can then be carried out by the binary long-addition algorithm.

$$\textbf{0101}$$
$$+\textbf{1011}$$
$$\overline{}$$
$$\textbf{(1)0000}$$

It seems to work; a proof is left as an exercise.

An implementation of this subtraction algorithm in digital logic gates is shown in Fig. 3.11. It invokes the functional block of Figs. 3.6 and 3.7 as a

Figure 3.11   Implementation of subtraction in digital logic.

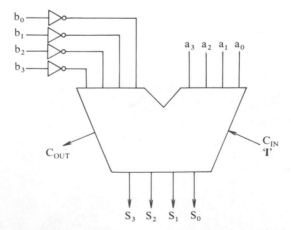

4-bit adder. The number **B**, $(b_0 \ldots b_3)$, is negated bit-by-bit in the four **NOT** gates. The addition of '**1**' needed to convert it into a negative number is performed by setting the carry-in bit of the adder to '**1**'. The circuit's operation can therefore be represented as

$$A - B = A +_{ar} \bar{B} +_{ar} C_{IN} \quad (1)$$

(The carry-out has special significance in this function; a study of it is left as an exercise.)

Now we add a little sophistication. The truth table for the **EOR** operation is presented in Fig. 3.12. Note that when the input **C** is low, the output is the same as input **B**, but, when **C** is true, the output is the *negation* of **B**. The array of Fig. 3.13 represents a multi-bit, controllable negation functional block. The input **C** is the *negate* control input. As a final stage of sophistication, this functional block could even be included in the multi-bit adder functional to yield a multi-bit add/subtract functional block (Fig.

Figure 3.12   The **EOR** gate as a controlled negation operation.

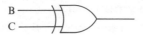

B	C	Output (B $\oplus$ C)	
0	0	0	$\left.\right\}$ C = 0, output = B
1	0	1	
0	1	1	$\left.\right\}$ C = 1, output = $\bar{B}$
1	1	0	

Figure 3.13   A multi-bit negation functional block.

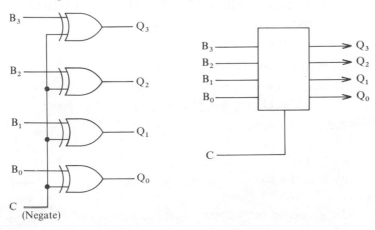

Figure 3.14   An ADD/SUBTRACT functional block.

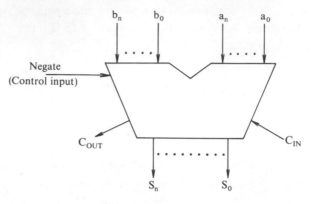

3.14). (The negation control input could be connected to the carry-in input but, as shown in the exercises, this restricts its versatility.) Most microprocessors incorporate a functional like this to carry out subtraction.

3.4   Multiplication with binary numbers

A simple algorithm for multiplication adds together the number of copies of the multiplicand specified by the multiplier. This is easy to organise in a machine, but the process is very slow for big numbers. The mechanical calculators in use before computers were developed refined this clumsy algorithm a little by using successive addition of a *place-shifted* multiplicand into an *accumulating register*.

Our common algorithm for long multiplication of decimal place-value numbers is quite efficient. It involves a rather complicated procedure using a memorised multiplication table and additions of partial products. In the table below, this procedure is depicted with a slight modification that requires only two partial products to be added at any stage.

$$
\begin{array}{rl}
105 & \text{(multiplicand)} \\
\times\,123 & \text{(multiplier)} \\
\hline
315 & = 3 \times 105 \text{ (partial product)} \\
+\,210\text{o} & = 2 \times 10 \times 105 \text{ (partial product)} \\
\hline
2415 & \text{first partial sum} \\
+\,105\text{oo} & = 1 \times 100 \times 105 \text{ (partial product)} \\
\hline
12915 & \text{second partial sum (final product)}
\end{array}
$$

This algorithm, with its big look-up tables for decimal multiplication and addition, is not very well suited to implementation in digital logic. However, the principles transfer very nicely into binary multiplication.

In binary arithmetic, the partial products require the multiplicand to be multiplied by two and then by the multiplier; both of these operations are trivial with binary numbers. First, consider multiplication by two. Since each place-value is weighted by a power of two, it is performed by a shift of each bit one place to the left (that is, to the next more significant place). Division by two corresponds to shifting all binary bits one place-value to the right; the remainder, the previous l.s.b., will 'fall-out' of the right-hand end of the multi-bit number.

Now we will consider the multiplication operation by a simple example,

$$10_{10} \times 5_{10}$$

In the long-multiplication algorithm form

$10_{10} = \mathbf{1010}$ multiplicand (shifts left)

$5_{10} = \mathbf{101}$ multiplier

$$
\begin{array}{r}
\mathbf{1010} \\
\mathbf{101} \\
\hline
\end{array}
$$

$$
\begin{array}{rl}
\mathbf{1010} & (\mathbf{1} \times \mathbf{1010}) \\
+ \ \mathbf{0000} & (\mathbf{0} \times \mathbf{10} \times \mathbf{1010}) \\
\hline
\mathbf{01010} & \\
+ \mathbf{1010} & (\mathbf{1} \times \mathbf{100} \times \mathbf{1010}) \\
\hline
\mathbf{110010} &
\end{array}
$$

Each partial product represents multiplication of the left-shifted multiplicand by the respective bit of the multiplier.

The look-up table for *binary* multiplication is trivial; if the multiplier bit is **1** the partial product is the current value of the multiplicand, if the multiplier bit is **0**, the partial product is **0**. The partial products are either the present multiplicand value or **0**. The simplification in changing from base-10 to base-2 is even more dramatic than the previous change from base-60 to base-10!

If the multiplier is right-shifted before each partial product is evaluated, the bit 'falling-out' can be used to control whether or not the present multiplicand value is to be added into the product.

Box 3.5
Multiplication using digital logic

At this stage, we will only outline the implementation of the algorithm; detailed analysis is left until Chapter 6. Fig. 3.15 shows the layout of the functional blocks required for long multiplication. The broad arrows at the adder are a common symbolic representation of a multi-bit digital connection.

The product is generated in an adder functional block which can be commanded to add the input word to its own output or contents. (An accumulating register or *accumulator*.) The binary number in the multiplicand can be commanded to shift one bit left. The binary number in the multiplier can be commanded to shift one bit right; l.s.b. 'falling-out' is used as a control input to the command generating logic.

The control logic implements the algorithm. If the multiplier right-shift outputs a logic **1**, the present multiplicand is added to the register; if a logic **0** is output, the multiplicand is simply not added. This is continued with left-shifts of the multiplicand until the process is complete.

This is the type of sequential operation that a computer must be able to perform. It is usually implemented in a microcomputer as a sequence of simpler logic operations. We will examine this in detail in Chapter 6 after we have described the detailed architecture and operation of the microcomputer.

Figure 3.15 A functional block layout for multiplication by the summation of partial products.

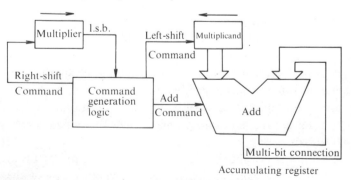

The sequence of logic operations for long multiplication will obviously take some time to perform. Thus, multiplication will be a slow process in comparison with addition and subtraction; it was a necessary 'trade-off' in the early days of microprocessor development. While it is acceptable for many software applications, it is a serious problem when very high speed arithmetic must be performed.

Multi-bit binary multiplication can be handled in a manner similar to that for addition (Box 3.3), as a combinatorial operation on the entire multiplicand and multiplier words. Let us represent a multi-digit word as bold capital symbols, e.g.

$$\mathbf{A}, \mathbf{B}$$

and represent the individual digits of a word by subscripted bold lower case symbols. Thus a generalised n-digit place-value word in the base $\mathbf{E}$ can be decomposed as

$$\mathbf{A} = \mathbf{a}_{n-1}\mathbf{E}^{n-1} + \cdots + \mathbf{a}_0\mathbf{E}^0$$

$$= \sum_i \mathbf{a}_i \mathbf{E}^i$$

Box 3.6 explores combinational multiplication for $\mathbf{E}$ equal to '2'.

Box 3.6
High-speed multiplication

Consider the multiplication of two unsigned binary multi-bit words $\mathbf{A}$ and $\mathbf{B}$, with the same number of bits, n. The application of the long multiplication algorithm described in the text can be written in general form thus

$$\mathbf{A} \times \mathbf{B} = \sum_i \mathbf{a}_i 2^i \mathbf{B}$$

$$\mathbf{P} = \sum_i \sum_j 2^{i+j} \mathbf{a}_i \mathbf{b}_j$$

Fig. 3.16 shows a tabulation of the product partials for $n = 3$. The individual product term truth table is simply the AND operation. The full set of $n \times n$ product terms can be generated by the combinational digital logic array of Fig. 3.17. The individual $n \times n$ terms must be added together to form the product P. Fig. 3.18(a) rearranges the partial sums so that they can be performed by the full-adder functional blocks explored in Box 3.2, with two argument words. The entire addition scheme for the multiplication algorithm can then be carried out by the logic circuit depicted in Fig. 3.18(*b*).

Some generalisations can be made about the combinational multiplier.

For n-bit words, the product will contain 2n bits.

For n-bit words the scheme depicted above will require n(n − 1) full adders.

The regular layout of the full adders suggests that a higher-level

Figure 3.16 A tabulation of partial terms for multiplication of two 3-digit binary numbers.

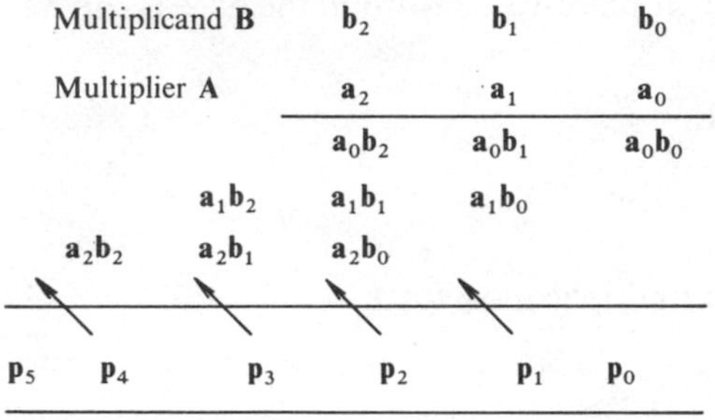

Multiplicand B
b_2 b_1 b_0
Multiplier A
a_2 a_1 a_0
a_0b_2 a_0b_1 a_0b_0
a_1b_2 a_1b_1 a_1b_0
a_2b_2 a_2b_1 a_2b_0
p_5 p_4 p_3 p_2 p_1 p_0

Product truth table

a_i	b_j	a_ib_j
0	0	0
0	1	0
1	0	0
1	1	1

Figure 3.17 Multiplier term array for Fig. 3.16.

b_2
b_1
b_0
a_2
a_1
a_0
a_2b_2
a_2b_1
a_2b_0
a_1b_2
a_1b_1
a_1b_0
a_0b_2
a_0b_1
a_0b_0

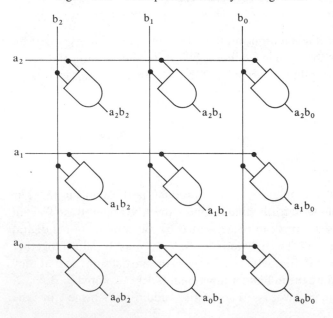

Figure 3.18 The scheme for using a full-adder array to implement multiplication of 3-bit words, (a) the partial sums and (b) the full-adder interconnection.

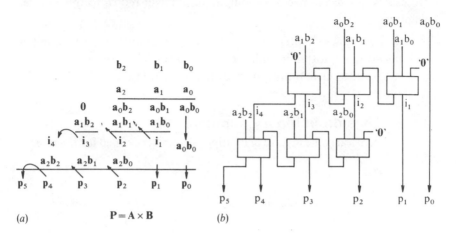

(a) $P = A \times B$ (b)

functional block could be developed to handle 'slices' of the scheme for horizontal, vertical or diagonal cuts through a symbolic diagram like that of Fig. 3.18(b).

The combinational multiplication scheme described in Box 3.6 does not handle two's-complement signed numbers. A generalisation to signed binary multiplication was devised by A. D. Booth in 1951. Since then, the *Booth algorithm* has been developed and modified to make it compact and fast. It is suitable for implementation in LSI functional blocks. Most popular logic families with LSI capabilities offer a multi-bit combinational multiplier. In the LS family, the 74LS274 allows 4×4-bit combinational multiplication with about 50 ns propagation delay, and the possibility of cascading to larger word sizes. In the *Advanced Low-power Schottky* (ALS) family, the 74ALS1616 LSI chip allows 16×16-bit multiplication in several selectable modes with propagation delays of about 50 ns.

The combinational multiplication algorithm can be developed to very high performance levels; the Hewlett-Packard HP1000 series computers use a single-chip 56×56-bit signed integer combinational multiplier with an overall propagation delay of less than 1 μs. This multiplier chip is one of a family designed specifically for very fast processing of arithmetic operations. The family uses CMOS silicon-on-sapphire technology.

A personal microcomputer seems very humble in comparison with such sophistication. Present trends indicate that such power will, however, soon be available in small computers. There is already considerable development of powerful arithmetic functional blocks (arithmetic co-processors) for use in personal computers. We will return to this in later chapters.

In the next few chapters we will use the simple binary number multiplication algorithm introduced at the start of this section, to outline the logic operations that must be performable within the architecture of a computer.

3.5   Division in binary numbers

Division is the inverse of multiplication; it is defined for the integer field as

$$B = Q \times A + R$$

infers

$$Q = B/A \text{ remainder } R$$

where

$$R < B$$

It can be carried out by the inverse of the simple multiplication algorithm, by repeated subtraction of **A** from **B** until the residue is less than **A**. The quotient **Q** is simply the number of subtractions carried out and the remainder **R** is the final residue. This tedious and slow process can be made fast and efficient by adapting the place-value *multiplication* algorithm described in Section 3.4. It requires the divisor first to be left-shifted until it aligns with the dividend. The number of times that the shifted divisor can be subtracted from (i.e. divided by) the dividend is determined and carried out before the divisor is right-shifted by one digit. This is repeated until the divisor has been restored to its original place. This awkward process is the basis of our common algorithms for long division.

The procedure for decimal integer division is depicted below.

```
            2 4 3
          _____
    27)6 5 8 1    align divisor and, by inspection, guess
     − 5 4 o o    the partial quotient, 2; remove 2 × 2700
       ___
       1 1 8 o
     − 1 0 8 o    by inspection, remove 4 × 270
       _____
         1 0 1
     −     8 1    remove 3 × 27
           ___
           2 0    remainder
```

The division algorithm is more complicated than multiplication because a test must be carried out at every stage. The initial left-shift steps must be stopped just before the divisor becomes bigger than the dividend; during the successive place-valued subtractions, the partial quotients must be (guessed and) tested to be less than, but within the divisor of, the residue.

Babbage's Analytic Engine and subsequent mechanical calculators determined the partial quotients by successive subtraction of the (shifted) divisor, counting the subtraction cycles until the dividend underflowed (went negative). One divisor value was then added back into the dividend before the divisor was right-shifted and the machine moved on to determine the next partial quotient.

You will be greatly relieved to find that this algorithm is much simpler in binary arithmetic. The simplification results, as in the case of multiplication, from the simplicity of the binary multiplication table.

$$\mathbf{B} \times \mathbf{a_i} = \mathbf{B} \qquad (\mathbf{a_i} = \mathbf{1})$$
$$= \mathbf{0} \qquad (\mathbf{a_i} = \mathbf{0})$$

There are thus only two cases to be considered; is the present dividend residue less than the present divisor, or is it greater than (or equal to) it? An obvious test is to subtract the divisor from the dividend and test if the result is negative. But this test yields the numerical difference; all that we need is a single bit of information about the comparison. The simpler operation of *arithmetic comparison* is explored in Box 3.7.

Box 3.7
Arithmetic comparison by digital logic

Simple combinatorial logic can test for equality of two multi-bit numbers (Fig. 3.19). Truth Table 3.1 depicts the three logic outputs required for full arithmetic comparison between two identical weight bits. Fig. 3.20 depicts a suitable circuit for its generation. Full magnitude

Figure 3.19   A logic circuit to test equality between two equal weight bits. The circuit can be cascaded as indicated.

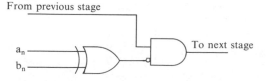

comparison between two unsigned binary multi-bit words is complicated somewhat by the need to account for the 'weight' of successive place-value bits. A full comparator can be built by extending and cascading the circuit of Fig. 3.20. Fig. 3.21 shows one stage of a circuit that accounts for the hierarchy of place-value weight. The block shown suffers from a long propagation delay time. For fast comparison, it is necessary to express and implement the outputs as direct combinational functions of all the bit and cascade-input variables. A study of this is left as an exercise.

Figure 3.20   A logic circuit for a 1-bit comparator.

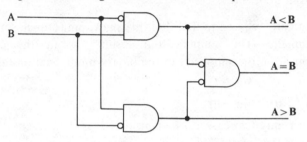

Figure 3.21   One stage of a multi-bit arithmetic comparator.

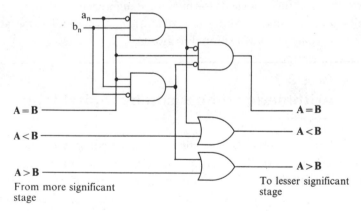

Truth Table 3.1

A	B	A = B	A > B	A < B
0	0	1	0	0
0	1	0	0	1
1	0	0	1	0
1	1	1	0	0

Fast multi-bit magnitude comparators are available in LS, TTL and CMOS logic families. The 74LS85 chip performs 4-bit magnitude comparison with a propagation delay of about 25 ns and can be cascaded for larger words.

Fast division combinational function blocks are available, but are not as common as the multiplier functional. They are usually included in powerful computers designed for fast arithmetic processing. Fast division hardware is not yet generally available in small microcomputers; division is usually performed by the binary long-division algorithm described above. We will explore this in Chapter 6 after computer architecture has been discussed.

3.6   Fixed- and floating-point binary numbers

A well-formulated theory of numbers and arithmetic is independent of the number base, and can be applied to binary numbers. A theory of numbers for human consumption can handle awkward concepts such as irrational numbers, infinity etc. by introducing new symbols such as

$$4/3 = 1.3333333\ldots = 1.\dot{3}$$

$$\infty, \quad \pi, \quad e, \quad 2^{1/2}, \quad -1^{1/2} \quad \text{etc.}$$

Some of these expanded concepts are not quite so easy to introduce into a *hardware model* in a machine. Our first example of this was seen in Section 3.3 when negative numbers and subtraction were introduced. An implicit method had to be used to avoid the need for a special symbol. The concept of binary point to mark the unit place-value encounters the same problem; there is no symbol available to mark its position. Once again, an implicit specification must be developed.

Another type of problem results from the finite nature of any hardware model. The continuous, real-number line can only be modelled by a set of discrete, separate points. The number of points, and their separation are intimately connected; Fig. 3.22(*a*) shows how different (implied) positions of the binary point generate different number patterns. The precision of numbers and arithmetic operations is *always* limited by the machine's hardware.

Within an 8-bit machine, a word can only have 256 states and can only specify that number of points on a number line. This is adequate for playing games and for a simple video display, but it is somewhat inadequate for comparing the size of an atom with that of the Milky Way galaxy.

Obviously, multiple words must be used to represent numbers. We have stated that a computer is simply a machine which carries out logic operations on multi-bit *words*. The word-size of the computer only limits the number of bits that can be handled *at the one time*. The machine must be designed such that *numbers* consisting of *multiple words* can be manipulated by a sequence of logic operations on individual words. We will explore this aspect of machine architecture at some length in Chapter 6.

The integer number line is modelled by assuming the binary point to be just to the right of the l.s.b. The common convention in microcomputer software is to use a signed integer of 16 bits; this gives a range of

$$-32\,768 \text{ to } +32\,767$$

The precision of the resulting number line is

$$1/65536 \quad \text{i.e. } 0.000\,15$$

The best achievable accuracy of the number representation is about $\pm 0.001\%$.

Higher precision and a longer number line can be readily achieved by using three or more words. Signed integers are conveniently modelled in two's complement by using the m.s.b. of the highest-order word as the sign bit.

The limitation of the fixed-point representation is not acceptable for a number model which must cope with a very wide range of magnitudes at high precision. An alternative representation is available in the well-known *exponent or scientific format* used in decimal calculators. In computer

Figure 3.22   Signed number lines for (*a*) 4-bit fixed-point two's-complement numbers with different binary point locations, and (*b*) the floating-point model with a 4-bit mantissa; only a few of the many possible exponent values are shown.

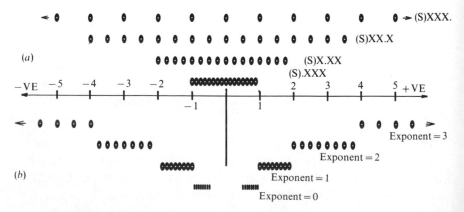

work, it is called *floating-point format*. A number is represented by a signed *mantissa* and an integer *exponent*. The number 1234.567 can be represented in any of the following forms.

$$1234.567_{10} = 1.234567 \times 10^3 \qquad (\text{exponent} = 3)$$
$$= 0.1234567 \times 10^4 \qquad (\text{exponent} = 4)$$
$$0.00001234567 \times 10^8 \qquad (\text{exponent} = 8)$$

The format can be applied to binary numbers simply by changing everything to base-2. The binary number **10110.101** can then be represented in any of the following forms.

$$\mathbf{10\,110.101 = 1.011\,010\,1 \times (10)^{100}} \qquad (\text{exponent } \mathbf{100}, \text{ i.e. } 4_{10})$$
$$\mathbf{= 0.101\,101\,01 \times (10)^{101}} \qquad (\text{exponent } \mathbf{101}, \text{ i.e. } 5_{10})$$
$$\mathbf{= 0.001\,011\,010\,1 \times (10)^{111}} \qquad (\text{exponent } \mathbf{111}, \text{ i.e. } 7_{10})$$

Whilst all of the representations above are equally valid the second allows the binary point placement problem to be solved neatly; the binary point can be implied just to the left of the m.s.b. of the mantissa. Furthermore, since the m.s.b. of the mantissa is always logic **1**, all bits of the word are used to establish its magnitude and it is represented with the best possible precision. If a convention is followed that the mantissa **M** of every floating-point number in the machine is such that

$$2^0 > \mathbf{M} \geqslant 2^{-1}$$

then the m.s.b. of the mantissa will always be **1** and will be redundant. That bit can therefore be replaced by a *sign flag* which indicates the sign of the mantissa in the conventional manner, **0** for positive and **1** for negative. (Note that the mantissa must always represent a magnitude; two's-complement representation cannot be used.) Fig. 3.22(*b*) shows the number lines represented for a small range of exponents.

The IEEE (Institute of Electrical and Electronic Engineers) standard format for floating-point numbers specifies a 32-bit mantissa for 'single precision' and 64 bits for 'double precision'. The Microsoft BASIC used in most personal computers uses a 4-byte (32-bit) mantissa with about 4×10^9 states. This provides a precision of about $\pm 10^{-10}$. It is invariably stored in signed magnitude form. The exponent is stored in a further byte with a range of ± 127; this yields a range of $2^{\pm 127}$ ($10^{\pm 30}$). To provide a simple representation of the number zero (see Box 3.8), the exponent is stored in *offset binary format*, where the origin is placed half-way along the number line. The origin of the exponent is thus (for an 8-bit exponent) at 128_{10}. This is called *excess-128_{10} format*.

Box 3.8
Floating-point number representation

This box demonstrates floating-point number representation within an 8-bit microcomputer. For simplicity, a single word is used for both mantissa and exponent. The mantissa is shown in signed magnitude form and the exponent in offset integer form.

Table 3.2 gives some examples which demonstrate the various features of the format. The full 8-bit binary mantissa needed for mathematical operations is shown in the table but, in the format used to store it in memory (*packed, floating-point format*), its m.s.b. is always masked by the sign flag as shown.

Table 3.2. *Some examples of floating-point numbers*

Decimal number	Binary number	Floating-point binary and (hex.)		Decimal equivalent
		Mantissa	Exponent	
1.0	1.0	^{1}(0)0000000 ($00)	10000001 ($84)	$\frac{1}{2} \times 2^1$
0.5	0.1	^{1}(0)0000000 ($00)	10000000 ($80)	$\frac{1}{2} \times 2^0$
0.75	0.11	^{1}(0)1000000 ($40)	10000000 ($80)	$\frac{3}{4} \times 2^0$
−0.75	−0.11	^{1}(1)1000000 ($C0)	10000000 ($80)	$-\frac{3}{4} \times 2^0$
−0.25	−0.01	^{1}(1)0000000 ($80)	01111111 ($7F)	$-\frac{1}{2} \times 2^{-1}$
8.5	1000.1	^{1}(0)0001000 ($08)	10000100 ($84)	$\frac{17}{32} \times 2^4$
0.0	0.0	^{X}XXXXXXXX	00000000	(see text)

The number zero cannot be modelled within the formalism; it requires an exponent of minus infinity. A special code must be assigned to it. The largest magnitude, negative exponent value is usually reserved to represent zero. (This represents the smallest number magnitude representable.) This is shown in the examples.

Floating-point format representation of binary numbers allows a very wide range of number magnitudes to be handled in a very compact form.

The format is not, however, suitable for direct application of arithmetic algorithms as discussed before for fixed-point numbers. The packed format used for number storage, with the m.s.b. of the mantissa masked by the sign flag, must be unpacked before arithmetic operations are performed and the sign flag must be handled separately.

Before a simple addition can be executed by the multi-bit addition functional blocks described previously, the two numbers must be adjusted to have the same *place-value* weights. This is done by right-shifting the lesser magnitude number and adjusting its exponent, until both numbers have the same exponent value (see Exercise 3.14 below). Negative numbers must then be converted to two's-complement format before the addition algorithm is used. The result must then be converted back into packed floating-point format. The procedure for multiplication and division is considerably simpler, but still requires the sign flag to be stripped before the actual arithmetic procedure is executed.

Floating-point format is a convenient, compact form for storage of numbers in a computer but, before the numbers can be used in arithmetic operations, they must be *unpacked* into their full, *expanded* format.

The architecture of the computer must be able to perform all of the operations described here. They can be performed by a sequence of simple logical operations on the computer variables or by sophisticated functional blocks. Chapter 6 will explore some examples of arithmetic operations as sequences of simple logic operations in a simple 8-bit microcomputer.

Powerful computers, designed principally for fast arithmetic operations, usually invoke VLSI functional blocks to handle all arithmetic operations directly on the multi-bit floating-point numbers; typical execution times are less than 1 μs. Powerful floating-point hardware functional blocks will soon be available in small computers; fast, arithmetic co-processors are already available in the newer machines. These will be described in later chapters.

3.7   Binary, octal, BCD and hexadecimal format

Multi-bit binary variables are very suitable as a model of binary numbers and arithmetic within an electronic machine, but they are not well suited to mere humans. Consider the two 16-bit binary numbers:

1010010010101110   1010001101010110

They are quite difficult to memorise and to compare. A more convenient representation is needed.

In the early days of computers these long binary numbers were grouped into blocks of three, starting from the l.s.b., and were represented by digits indicating the binary weighted values (0–8). This is the octal representation (base-8) of binary numbers. The two multi-bit examples above are represented thus:

$$122256_8 \qquad 121526_8$$

It is still in use but is very awkward when dealing with 8-bit or 16-bit words. If the two numbers above are both split into two 8-bit words, the results are

$$244_8, 256_8 \qquad 243_8, 126_8$$

and they are difficult to relate to their 16-bit representation. Since microcomputers are now solidly based on multiple words of 8 bits, this awkward system has been replaced with hexadecimal representation (base-16). Within this system, the 8-bit words decompose neatly into two groups of 4. There are not enough decimal numerals to represent each 4-bit state; alphabetic symbols are used for the last six. The sixteen states are symbolised as shown in Table 3.3. To avoid the need for a subscript, the hexadecimal representation is usually indicated either by the letter H at its end or by the $ symbol at its beginning. The two numbers above are thus coded as

$A4AE \qquad $A356

A4AEH \qquad A356H

(The prior form will be used throughout this book.)

The two 16-bit binary words above (four hexadecimal digits) can be decomposed simply into two 8-bit binary (two hexadecimal digits) words:

$A4, $AE \qquad $A3, $56

Nicknames are given to the various bundles of bits. The name *bit* is itself a

Table 3.3

Decimal	0	1	2	3	4	5	6	7
Binary	0000	0001	0010	0011	0100	0101	0110	0111
Hex.	0	1	2	3	4	5	6	7
Decimal	8	9	10	11	12	13	14	15
Binary	1000	1001	1010	1011	1100	1101	1110	1111
Hex.	8	9	A	B	C	D	E	F

contraction of *binary digit*. The group of four bits representing a hexadecimal digit is commonly called a *nibble* or *nybble*, probably because a group of 8 bits is called a *byte*. The nybble and byte were of special significance before binary machines became popular. Four binary bits are needed to specify the ten states of a decimal digit (see Table 3.3). BCD numbers were extensively used for decimal machines, and words were in multiples of 4 bits. (Refer to the early IBM BCD text code described in Box 2.4.) The name *word* is usually reserved for the number of variable bits handled simultaneously as data in a computer.

Closing remarks

This chapter has explored the power and efficiency of binary logic as a framework for numbers and arithmetic operations. We have seen that binary logic can model all the basic arithmetic operations subject to the limitations of precision, time and space imposed by the need to carry out the procedures in real, finite machinery.

Many of the operations require execution of a sequence of simple logic operations with the storage of intermediate results. In the following chapters we will see how computer hardware can be organised within an architecture which allows virtually any arithmetic or logic task to be performed as a sequence of simple binary logic operations.

Exercises

3.1 (*a*) Convert the following binary patterns to hexadecimal representation.

   (i) **00011001**

   (ii) **10000001**

   (iii) **10101011**

   (iv) **1110101**

   (v) **1000100110101011**

(*b*) Convert the following hexadecimal codes into binary patterns.

   (i) $11

   (ii) $0F

   (iii) $DC

   (iv) $ACDC

   (v) $1234

(*c*) Convert the following decimal numbers to binary unsigned integer representation. (The remainders of successive divisions by 2 will generate the binary number.)

(i) 17

(ii) 100

(iii) 1000

(iv) 170

(v) 65365

3.2 Carry out the unsigned integer additions below in binary arithmetic, then check your results by converting the addend, augend and sum to decimal representation.

(*a*) $1010 +_{ar} 10100$

(*b*) $10101 +_{ar} 1010$

(*c*) $1011 +_{ar} 10101$

(*d*) $101111 +_{ar} 110111$

3.3 (*a*) If you have access to a digital logic demonstrator, wire up the logic diagram of Fig. 3.3 and check its truth table. (If your logic set has only **NAND** or **NOR** gates you must first convert the circuit to these gates alone.)

(*b*) Repeat (*a*) using Fig. 3.4.

(*c*) Repeat (*a*) using Fig. 3.5.

3.4 (*a*) Consider the addition of two, 2-bit, positive binary numbers **A, B** with a carry input C_{IN} into a 3-bit sum.

Draw up a truth table and, from this, write down the boolean equations for the three outputs (s_0, s_1, C_{OUT}) in explicit terms of the five inputs $(a_0, a_1, b_0, b_1, C_{IN})$. Reduce these equations where possible to develop a manageable functional block for this 2-bit full adder with *look-ahead carry*.

If you have a logic demonstrator, express your boolean equations in available logic functions and connect them up to test their performance.

(*b*) Expand your solution to 3-bit addition. (Hint, see Fig. 3.8.)

3.5 (*a*) Prove that

$$\bar{A} +_{ar} 1 = -(A)$$

(Note that the numbers represented by **A** are ordered and equispaced integers; hence if one transpose can be proven, all are valid.)

(*b*) Show that the addition of a binary number to its bit-wise negation results in all bits being set. Use this result to prove the subtraction algorithm which uses two's-complement binary numbers.

3.6 Convert the following negative decimal numbers to an 8-bit two's-complement binary representation.

(*a*) -1

(*b*) -15

(*c*) -64

(d) −96

(d) −127

(e) −128

(f) −129

(g) −1.875

(h) −0.03125 (1/32)

(i) −73/128

3.7 (Important.) Read Box 3.4 and then consider the role of the carry input in two's-complement representation of subtraction (actually the addition of a negative number). By modelling the operation of a multi-bit adder as described, examine the effect of the carry input being true or false. You will find that it behaves like the negation of the *borrow* commonly used in the long subtraction algorithm.

Use this result to consider the behaviour of the carry output of an adder functional when the result should be positive (no borrow) or negative (borrow).

Check that the borrow in and borrow out for subtraction are simply the negation of the carry in and carry out used for addition.

3.8 Using the multiplication algorithm for binary numbers from Chapter 3.4, show that the product of an n-bit multiplier and an m-bit multiplicand will have up to $m+n$ bits.

3.9 Using the algorithm for unsigned binary multiplication, carry out the multiplications below. Check your result by converting all to decimal numbers.

(a) **1010 × 1010**

(b) **11 × 10001**

(c) **111 × 1111**

(d) **1010 × 101**

(e) **10.10 × 01.11**

(f) **0.0110 × 0.1001**

3.10 The 'Russian peasants' algorithm' for long multiplication requires only addition and division/multiplication by two. It involves repeated application of the following simple steps.
– If the multiplier is odd, the multiplicand is accumulated into the product.
– The multiplier is then halved and the multiplicand is doubled.
The process is repeated until the multiplier becomes 1.

Show that this is the algorithm described in Section 3.4, but executed on decimal numbers.

3.11 By exploring the products of some 4-bit signed binary numbers, devise a general algorithm for signed multiplication of two's-complement binary numbers.

3.12 Consider the multiplication of two 2-bit numbers as a combinatorial logic problem. Write down the 16-row truth table for the four output bits and from this derive boolean functions for each output in terms of the four inputs. If you have a logic demonstrator, connect up and test your two-bit multiply functional.

3.13 Examine the possibility of 'slicing' the full adder array in Fig. 3.18 (Box 3.6). Design a fast functional which can be 'cascaded' to perform the full scheme.

3.14 Design a combinatorial circuit which compares two 3-bit numbers and generates arithmetic comparison outputs.

3.15 Convert the decimal fraction $\frac{1}{3}$ into a binary number. (Divide the binary representation of 1 by the binary representation of 3.)

3.16 Using the simplified model described in Box 3.8, unpack the floating-point numbers specified below (signed magnitude mantissa first, offset exponent second) into a format suitable for the operation indicated. Carry out the operation and then repack the result. Check your answers by converting all to decimal numbers.

(*a*) **(01000000, 10000000)** $+_{ar}$ **(00000000, 10000001)**

(*b*) **(01000000, 10000000)** $+_{ar}$ **(00000000, 10001000)**

(*c*) **(01000000, 10000000)** $-$ **(00000000, 10000001)**

(*d*) **(01000000, 10000000)** $\times$ **(00000000, 10000001)**

(*e*) **(01000000, 10000000)** $\times$ **(10000000, 10000001)**

(*f*) **(1000010000, 10000000)/(10000000, 10000001)**

References and further reading

This list is only an entry into the extensive literature on this subject.

1 *Computer Design Development – Principal Papers*, E. W. Swartzlander (ed.), Hayden, New Jersey (1976)
   A collection of several important papers on computer and (especially) arithmetic algorithm development.
2 'A Signed Binary Multiplication Technique', A. D. Booth, *Quarterly Journal of Applied Mathematics*, **4**, part 2, 1951, pp. 236–40
3 'The IBM System/360 Floating Point Execution Unit', S. F. Andersen *et al.*, *IBM Journal of Research and Development*, **11**, 1967, pp. 34–53
4 'Floating Point Chip Set Speeds Real-Time Computer Operation', W. H. McAlister & J. R. Carlson, *Hewlett-Packard Journal*, **35**, No. 2, 1984, pp. 17–34
5 *Computer Architecture and Organisation* (ibid. Chapter 1)

4

Computer data transactions

Preamble

The previous chapters have described digital electronic logic and have shown how it can be used for control, to represent changing states of a system and to model numbers and mathematical operations. All of this is achieved by simple, binary, logic operations on multi-bit binary variables. It should now be obvious that a universal machine to carry out all such tasks, a computer, only needs to execute predefined sequences of logic operations on multi-bit binary variables; the apparent complexity of its operation derives from a clever arrangement or *architecture* based on simple logic operations.

There are several functions that we must understand, such as the logic circuitry which carries out the range of logical operations on data words, the system which allows the multi-bit variables to be transported back and forth between the other blocks, and the circuitry needed to load data and instructions into the machine and to output the results.

This chapter will describe the logic circuits which allow a specific multi-bit variable to be selected from amongst its many peers in a memory array, to be transported about within the machine or to be moved back into memory for storage. This type of function is called a *data transaction*.

4.1   Information transfer networks

A computer must have a communication system which can transfer multi-bit words between its various elements, in particular, between its *logic processor* where multi-bit logic operations are executed, and the memory elements where the words are stored. A common type of communication system is the *star network* of an automatic telephone exchange. Every *slave element* (subscriber) has a unique connection to the *master element* (the exchange). In a computer, the slave and master are the processor and the memory elements respectively. If there are several masters (as in the linked network of an automatic telephone system), the interconnection mesh becomes very complex. This is required in a telephone system to allow many transactions to take place simultaneously; each slave element needs a discrete pathway to the master. In a microcomputer, where there can be several hundred thousand separate memory cells each containing 8 or 16 bits of binary data, a star network would require hundreds of thousands of individual connections between the elements; this is obviously impracticable.

A party-line network is a simpler alternative (Fig. 4.1). Its simplicity comes at the expense of two disadvantages which can be demonstrated with a party-line telephone system.

(*a*) Only one transaction (conversation) can take place at a time.

(*b*) All slave elements (subscribers) must respond to every call to ensure that the required slave undertakes the transaction.

As we will soon show, neither of these are significant disadvantages for master/slave communications within contemporary (von Neumann architecture) microcomputers; only one transaction needs to be carried out at any one time between the master and a slave.

The shared party-line concept is applied universally in computers as a *bus system*. The name bus derives from Latin, *omniBUS*, belonging to everyone (hence *motor omnibus*).

The multi-bit variable transactions of a computer are carried out over a multi-wire party-line network called a *databus*. This uses a separate wire for each logic signal line of the multi-bit variable word; each signal line connects to the corresponding bit in every master or slave element within the computer. It is arranged such that data can be transferred from or to any element (bidirectional transactions).

Figure 4.1   The interconnection scheme of a party-line network.

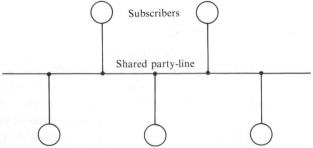

It is obvious that absolute pandemonium will result unless a very strict, well-defined procedure or *protocol* is defined for the orderly management of these party-line data transactions. We will now examine this *data transaction protocol*; we will start with the actual communication of a multi-bit word across the data bus, we will then examine the manner in which the source and destination of the data word are defined and we will finish by describing how the entire data transaction process is controlled.

4.2   The databus

The databus is a set of party-lines that connects together the corresponding bits of every multi-bit element in the computer (Fig. 4.2). Since the databus lines go to every element, the lines are of necessity quite long and have considerable inductance and capacitance; when logic levels are driven onto them, a short settling time is required before the logic levels become stable and correct.

At any one time, only one element of the computer can be allowed to impose logic levels on the databus lines. This element is said to *write data* onto the bus. It invariably writes all bits of the word simultaneously. After the necessary settling time, the element which requires the data can latch (*read data*) from the bus. This simple sequence allows a single multi-bit word of data to be transferred from one element to another in a relatively short time. The latched data is stored at the reader and the bus is then freed for subsequent transactions between other elements. (Box 2.2 demonstrated this time-multiplexed operation; the latching function used for reading the data was described in Chapter 2.2.)

Figure 4.2   The interconnection scheme of the databus.

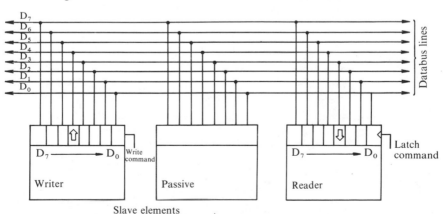

The logic gates that were described in Chapter 1 are not suitable for writing data to the bus; standard TTL, CMOS, etc., gates do not allow logic outputs to be connected together, whereas a bus system specifically requires this. Obviously, different logic gates are required for writing to the bus; these are called *bus driver gates*. Some examples are described in Box 4.1.

Box 4.1
Bus logic systems, drivers and receivers

Ordinary TTL or CMOS families do not allow gate outputs to be tied together; if this is done and the tied gates attempt to impose different logic levels on the one line, the result will be indeterminate. This is quite unsuitable for a bus system where many gates must be able to write a logic signal onto the common lines.

Since the development of digital electronic logic, there has been a need for logic signal lines which could be driven by the output of several gates. A modified form of the standard gate output was developed for this purpose; it was called the *wired-or* output. It is easily achieved in TTL and its derivatives by an *open-collector output stage*. The modified output is shown in Fig. 4.3 (cf. Fig. 1.7). The transistor in the TTL gate which pulls the output voltage up is omitted! The gate is thus only able to generate a logic low output when it sinks current through its single output transistor to

Figure 4.3   Wired-or, open-collector gates, (*a*) internal circuitry, (*b*) symbolic representations.

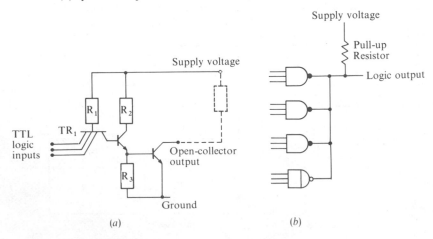

ground. If several of these gates are tied together and connected to the positive supply line through a resistor, then, if all of the output transistors are turned off, the external *pull-up resistor* will lift the output to logic high, but if any output transistor is turned on, the output goes to logic low. A common symbolic representation for this type of output is to show it with a filled negation ball. (An alternative form is also shown in the figure.) A limited number of gates can thus drive a single line, but fan-out is limited by the current capability of the output transistor and by the input current requirement of the driven TTL gates. Since the high logic level current is provided by the *passive pull-up* resistor and not by the *active pull-up* transistor of a normal gate output, the rise-time from logic 0 to logic 1 of the common line is determined by the time constant of the pull-up resistor and the output line capacitance.

Open-collector lines are restricted to non-critical control signal lines in microcomputers; their fan-out limitation and slow response precludes their use in general databus applications.

Microcomputer bus logic usually employs the LSTTL *three-state logic* or *tri-state logic* family. (Tri-state is a proprietary name used by National Semiconductors.) The normal output stage of LSTTL is rearranged so that it can be selectively *disconnected* from the bus lines. This is achieved by adding a control line to divert the base currents away from the output stage transistors, thereby preventing them from driving the output to either logic 0 or 1; when this is done, the gate is said to be *disabled* (Fig. 4.4). This type of gate makes it possible to have just one of the many output gates connected to a common bus line *enabled* (or active) at any one time. The output transistors of the selected gate can force the line quickly to logic 1 or 0. The names *three-state* or *tri-state* obviously refers to the number of states that the output can adopt, logic 0, logic 1 or the third state, commonly called the *high-impedance or Hi-Z state*. The output maintains the same logic levels but has a higher current capacity than the standard LSTTL gate; it can provide line driving currents of the order of 20 mA in both active states. This allows long, heavily loaded and high-capacitance bus lines to be driven reliably with short settling times.

Fan-out on three-state busses is increased enormously by the use of specialised *bus reader gates*. These have PNP transistors ahead of the normal logic input structure; this reduces the gate input current to about 0.03 mA and allows a notional fan-out of several hundred.

The 74LS373 and 74LS374 devices are typical bus driver/receiver gates. Both have the three-state output and the low current inputs needed for bus

line driving and reading respectively. Inputs and outputs are both LSTTL
compatible. The 74LS373 is an 8-bit transparent D-latch; the 74LS374 is
an edge-triggered 8-bit D-latch. Both have a single ENABLE control line
which forces the 8-bit latch contents onto the eight three-state outputs (see
Fig. 4.5). Bidirectional bus reader/writer gates are also available (Fig. 4.6).
Refer to data manuals for the 74LS240/241/244/245, the 74LS640-5 and
the 74LS646-9 gates. We will see in later chapters that these devices are very
convenient for both writing and reading from 8-bit data busses.

The three-state system is very convenient for microcomputer busses where
the lines are fairly short; there is, of course, a limit to its application. The
settling time is determined by the effective capacitance and inductance of
the line and by the number of gates reading it. The bus structure acts like a
transmission line and logic level changes propagate along it; reflections
from the unterminated ends are damped only by the impedances of the
readers and of *enabled* bus drivers. With long lines and unsuitable loads,
bus transients and reflections can take a long time to decay; this seriously
affects the settling time. Care must be taken with the layout and loading of
the bus structure to minimise these effects in big systems.

Figure 4.4   (*a*) The general scheme for disabling a three-state output
gate and (*b*) its symbolic representation.

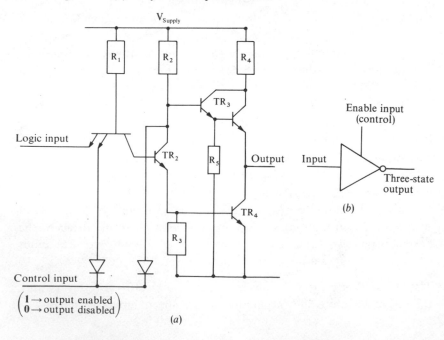

When the bus lines must be long and the maximum possible data transaction rate is required within the system, the bus must be made from properly terminated transmission lines. Fig. 4.7 shows a typical configuration for a 300 Ω terminated bus system. Both ends of each bus line are terminated by two resistors which hold the bus line at an intermediate voltage (or at a voltage safely within the high logic level) when it is unloaded, and give an a.c. termination impedance of 300 Ω. This allows rapid settling of the line to logic **1** or **0** and cancels reflections at both ends. Three-state bus driver gates (or specially designed open-collector gates) are suitable for driving these busses; but receivers must all have high-impedance inputs.

With careful design, a terminated-line bus system will be limited only by the propagation time of signals along the lines; with a suitable *handshake* control system (see Box 4.4) and repeaters to recover logic levels from induced noise, the bus length can be extended indefinitely. The Digital Equipment Corporation DEC PDP-11 minicomputer family, first introduced in 1969, used such a system.

Figure 4.5 The 74LS374 bus driver/latch; (*a*) functional layout; (*b*) typical symbol.

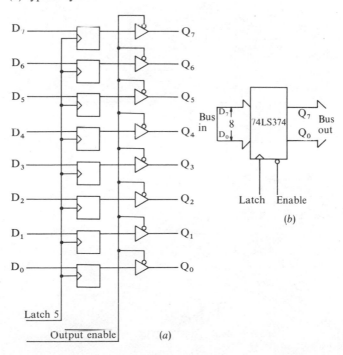

All of these bus systems require the return current to flow in the system ground; they are all therefore susceptible to ground signal noise. If the line capacitance is very high and the system must switch very fast, the ground currents can be of the order of amperes. Balanced pair, differential bus systems eliminate this. The reader is referred to the data manuals on *differential bus drivers* and on the Emitter Coupled Logic (ECL) family. (See also Chapter 8.)

Figure 4.6   The 74LS245 transceiver; (*a*) functional diagram and (*b*) its symbol.

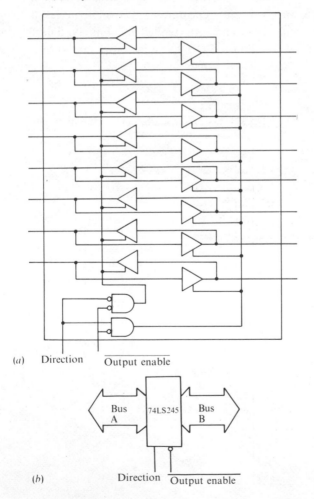

(*a*)   Direction $\overline{\text{Output enable}}$

(*b*)   Direction $\overline{\text{Output enable}}$

Figure 4.7   A terminated transmission line bus system.

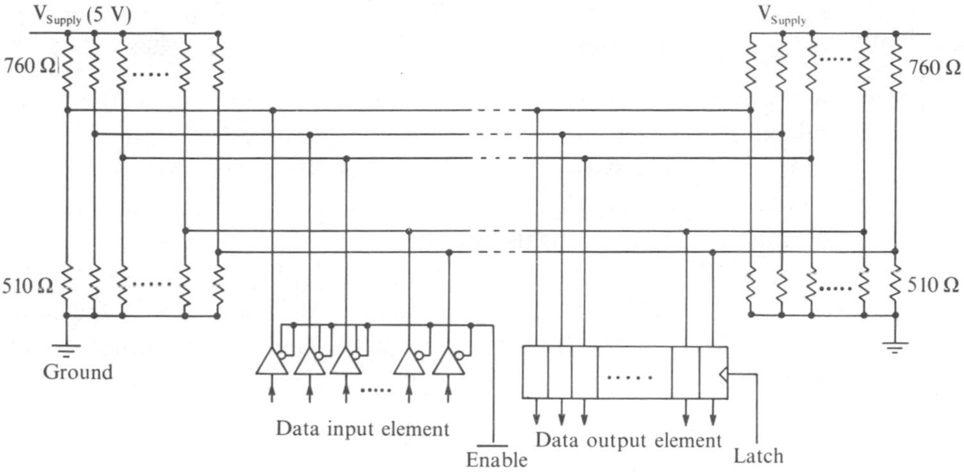

4.3   The address bus

This section will describe how a single reader or writer element can be selected from the many connected to the bus.

When a party-line telephone network was being described at the start of this chapter, it was noted that all elements of the system had to respond to every call so that the nominated element could be selected. In a computer bus system, the nomination of the selected elements is not made over the data bus lines, a second multi-line bus system called the *address bus* is used instead. It is connected to all slave elements and each is assigned a unique predefined code or pattern called an *address*.

The element of the computer in control of the transactions is the *bus-master*; the elements that it controls are the *slave elements*. The master designates the slave involved in the transactions by placing the code or pattern assigned to the designated slave on the address bus. All slaves must watch this bus and must engage in a transaction only when their own unique code or *address* is present. In a microcomputer, the microprocessor is usually the master of the bus and invariably is involved in every transaction as either the writer or reader of data; it only needs to use the address bus to designate one single slave element to act as either the reader or writer respectively.

Every possible bit pattern of the address bus can be used to designate a different element. In microcomputers with an 8-bit data word, the address bus usually has 16 lines so that 2^{16} (64K where $K = 1024 = 2^{10}$) individual addresses or elements can be designated. The more powerful 16-bit data word microcomputers have 20- or 24-bit address busses and can therefore designate 1M or 16M different elements on the bus ($M = 1048576 = 2^{20}$).

With a 16-bit address bus, it would seem at first sight that each of the 65 536 individual slave elements would need a 16-bit address decoder to recognise its unique address. We will now show how the decoder requirement is made a little more reasonable. Most of the address space in a computer is assigned to memory elements. Mass memory circuits are now at the stage where 256K binary memory cells can be fitted onto one silicon chip (Section 2.4 and Chapter 8). We will describe the application of the more modest 16K memory chips, to demonstrate some important features of economical address decoding schemes.

The common 16K memory chips are organised as a matrix of 128×128 individual 1-bit cells. Eight chips are arranged in parallel to provide 16K 8-bit memory elements, each chip represents one bit of the 8-bit computer word and is connected to the corresponding line of the databus. The depiction of the bit-cell matrix in Fig. 4.8 shows how the chip's decoding scheme uses 14 address bits to specify each individual cell. (A further description is given in Section 8.3.) Two 7–128 line decoders are included in the drive logic circuitry of the memory chip; in other words, all but the two most significant bits of the computer's 16 address lines are brought to each

Figure 4.8   16K memory chip decoding scheme.

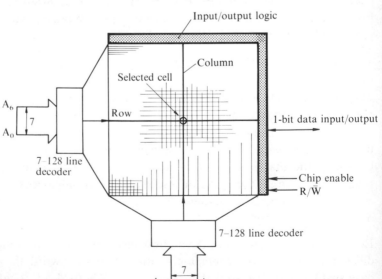

of the memory chips for decoding therein. The remaining two address lines are decoded in a 2–4 line decoder to divide the address space into four logical blocks or quadrants (Fig. 4.9). The outputs of this decoder are sent to the ganged CHIP ENABLE input of each row of memory chips; the CHIP ENABLE input allows the chip to be selected for transactions with the databus. Thus, the first three quadrants of the address space are fully decoded to the 48K memory elements. The remaining 16K of the address space could be assigned to other types of memory and to input/output addresses; this will be discussed further in subsequent chapters.

4.4 Transaction timing and the control bus

The concepts and hardware of a bus system have now been described in sufficient detail to allow the overall procedure or protocol for a data transaction between two elements to be described. It is obvious that the data transaction procedure must be managed by a logic system which controls the data and address bus systems described so far; this extra logic system is usually called the *control bus* of the machine.

We will consider a simple but complete bus system modelled on that used in simple 8-bit microprocessors, but reduced to its bare essentials. It uses only logic circuits which have been described previously.

Figure 4.9 Address and data logic layout for 48K × 8-bit memory using 16K × 1-bit chips.

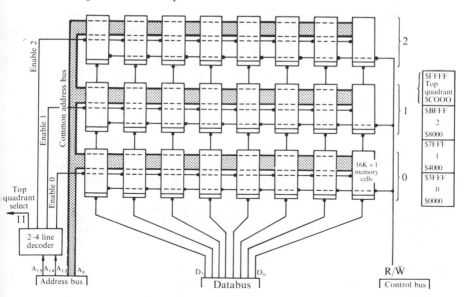

Consider the bus system depicted in Fig. 4.10(*a*). It has a single master element and four slave elements. Each element has a 4-bit data latch and a 4-bit driver connected to the databus; each is thus able to write to, or to read from, the databus. The master is able to write a 2-bit pattern onto the 2-bit address bus and each slave contains a decoder connected to this bus. Each decoder is a simple comparator which generates a logic **true** output in response to a predefined unique pattern being present on the address bus.

Figure 4.10   A rudimentary bus system; (*a*) block layout, and (*b*) its schematic logic diagram.

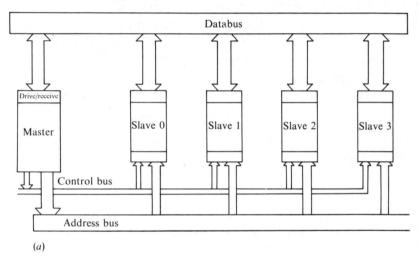

(*a*)

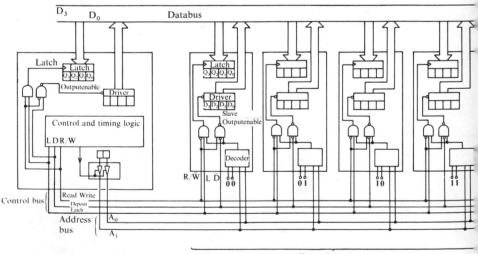

(*b*)

The master element contains control and timing logic to manage the data transaction. The transaction requires a sequence of actions called a *bus transaction cycle* to be carried out. The transaction sequence consists of three well-defined *phases* (see Fig. 4.11(*a*)).

(1) The master must first deposit the code of the designated slave onto the address bus. The address bus lines take some time to settle to the correct pattern; during this time the decoder outputs in the

Figure 4.11   (*a*) Basic data transaction cycle timing for Fig. 4.10. (*b*) The complete signal sequence on all three busses during two full data transaction cycles. In the first cycle, a 4-bit word of data is transferred from the master to slave #3 (binary address code **11**); in the second, a 4-bit data word is transferred from slave #1 (**01**) to the master.

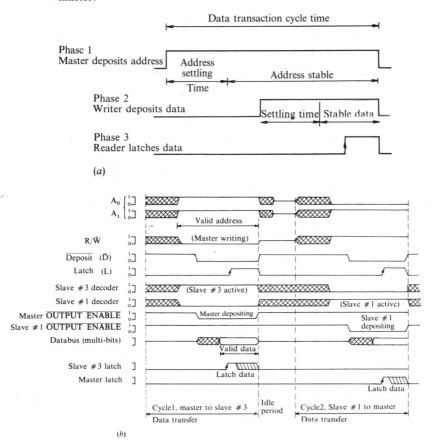

(*a*)

(*b*)

slaves are indeterminate and nothing can be done. When the address bus levels are stable, only the designated slave will have its decoder output asserted **true**.

(2) The second phase can begin as soon as the first is complete; the element which is to write data must now deposit it onto the databus. One of the logic lines of the control bus is used to signal to all elements whether the master or the slave is to deposit (write) data to the databus. In the figures, this line is labelled by its conventional name, the $READ/WRITE$ *line* (R/$\overline{\text{W}}$). When it is **true**, it indicates that the master is to read from the databus, when **false**, the master is to write to the databus (the truth value relates to the master). The control logic of the master determines when this second phase must start by driving the $\overline{DEPOSIT}$ ($\overline{\text{D}}$) line in the control bus low.

The logic circuits in both master and slaves use the $\overline{\text{D}}$ and the R/$\overline{\text{W}}$ signals to generate the required databus driver enable signal in either the master or the designated slave. Data is thus written onto the databus. A brief settling time is again required to allow the databus to become stable.

(3) The third and last phase of the transaction cycle can then be carried out. The valid data present on the databus is simply latched into the required element. The R/$\overline{\text{W}}$ line is now used to determine whether the master or the slave element is to latch, and the LATCH signal line of the control bus is used for timing the latching operation.

The logic signals generated by the master, to time and control the three distinct steps or phases of this transaction sequence are represented in their simplest form in the timing diagram shown in Fig. 4.11(*a*). Several loose conventions which are used to represent the states of three-state and bus lines are shown in Fig. 4.11. The two address lines at the top of the diagram show how the settling time is depicted (hatched regions), and how a three-state line is depicted when no gate is driving it (level midway between **0** and **1**). The multiple lines shown on the falling edge of the LATCH signals are a conventional method for showing that the timing of the falling edge is not important. The databus representation lower in the diagram depicts how a multi-bit information on a bus is often shown as one *trace* when the logic levels are not significant to the operation being explained; when the bus is not being driven it is depicted as for a single line (level midway between **0** and **1**), when it is settling it is shown hatched and when the levels are stable it is shown as two lines corresponding to logic **1** and **0**.

Box 4.2
Memory elements on a bus system

The simple bus model depicted in Figs. 4.10 and 4.11 has simple multi-bit digital logic inputs and outputs at every element; the aim is only to demonstrate data transactions. Other concepts of computer operations can, however, be understood from this model if some other features are added. Imagine that a set of manual switches are added to the master element to provide logic inputs to its bus driver, and that a set of indicator lights are added to the master's data latch outputs. If the data outputs and inputs are connected together bit for bit at a slave element, data latched into that slave can be read back by the master. This models the behaviour of a multi-bit R/W memory element on the bus.

Bus transaction demonstration boards with all of these features are used for laboratory work in the courses from which this book has evolved. They cycle through the full transaction sequence very slowly so that indicator lights on the data busses etc. can be watched to see the sequence take place. The multi-bit outputs and inputs on the slave elements can be connected together to imitate memory elements. Alternatively, they can be connected to logic input switches and logic state indicators, or to multi-bit logic function blocks. These allow multi-bit **AND**, **EOR** operations or arithmetic addition to be performed between pairs of multi-bit variables *programmed* in by hand, at the master input.

The bus transaction protocol discussed above is very like that implemented in 8-bit microcomputers. Section 4.5 describes the data transaction protocol and timing used with two common 8-bit microprocessors.

4.4.1   Handshake protocols in bus transactions

This section describes some of the powerful capabilities that can be provided in more sophisticated bus systems.

4.4.1.1   *Master–slave handshakes*

A large, sophisticated bus system usually includes many different types of slave elements which have widely differing response times. Memory elements are a typical example of this; different types of integrated circuit memory chips have response times ranging from 15 ns to 500 ns.

Other slave elements can have much slower response times. A system using such a range of elements can only work at maximum possible speed if the data transaction cycle time is varied to suit each active element. This cannot be achieved with the transaction protocol described so far, since there is no provision for a *feedback* signal from the designated slave; the entire system must work at the speed of the slowest element. The addition of a feature commonly called a *handshake* provides the feedback required. This allows every data transaction to be executed at maximum possible speed for many elements whose response times vary considerably. The data transaction protocol of sophisticated computers can, furthermore, provide a system which allows several master elements to share, and to compete for, access to a single set of busses. We will only outline the principles involved; they are not actually needed at this stage, but it is a logically convenient point to introduce them. They are, however, very important for powerful data-collection systems; some examples of their application will be described in later chapters.

We deal first with the problem of varying response times of slave elements. This is usually handled by adding a handshake to the data transaction protocol. Extra command signals issued by the master and acknowledged by the slave, are used to time the transaction through its various phases. A typical procedure for both a master/slave and for a slave/master transaction is described below and is depicted as a timing diagram in Fig. 4.12; it is a little more elaborate than the simple procedure of Figs. 4.10 and 4.11.

(*a*) Master-to-slave data transaction
  (1) The master starts the transaction by depositing the address, data and $R/\bar{W}$ signals on the respective busses; it waits for a short time to allow for settling time and for the designated slave's address decoder to detect its pattern.
  (2) The master then delivers a synchronising signal over a control bus line to command the designated slave to respond. This signal is commonly called the MASTER SYNC.
  (3) The designated slave then responds to the MASTER SYNC. and $R/\bar{W}$ signals by latching from the databus.
  (4) When the slave has completed its latching, it responds with another synchronising signal on the control bus, commonly called the SLAVE SYNC.
  (5) When the master receives the SLAVE SYNC. signal it immediately removes its own MASTER SYNC. signal.
  (6) The slave then responds to the removal of the MASTER SYNC. by removing its own SLAVE SYNC., marking the end of the cycle.

Note that the master can begin to set up the next transaction as soon as it receives the SLAVE SYNC. response, but that it cannot deliver the following MASTER SYNC. until the previous SLAVE SYNC. is released. (This allows the next transaction to start before the present one is finished; it is one form of *pipelining*, see Chapter 8.)

(b) Slave-to-master transaction
  (1) The master first deposits the slave address and the R/$\overline{\text{W}}$ signals.
  (2) The master then waits for the settling and decoding time before issuing the MASTER SYNC. signal.
  (3) When the designated slave receives the MASTER SYNC., it notes that it is to write, and then carries out internal operations to get the data ready. When ready, it deposits onto the databus and it then signals on the SLAVE SYNC. line.

Figure 4.12 Simple handshake protocol; (a) the control signals, and (b) the interactive timing.

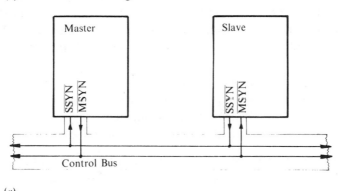

(a)

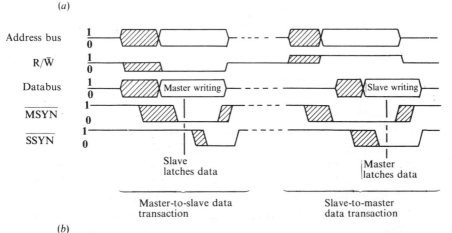

(b)

(4) When the master receives the SLAVE SYNC. signal, the databus will already be stable. It latches the data and removes its MASTER SYNC. signal.

(5) When the MASTER SYNC. is removed, the slave responds in turn by removing its SLAVE SYNC.

Note that the removal of the SLAVE SYNC. again signals the end of the transaction, and that the cycle will always be performed in the shortest possible time no matter how slow the slave is.

The protocols described for data transactions in either direction ensure that communication can be made reliably over very long lines; line settling and propagation delay times are accommodated automatically. If the master addresses a non-existent slave, it will never respond and the system would *hang-up* forever. This is usually avoided by adding a *time-out detector* which is activated to divert the system to some other operation if a SLAVE SYNC. signal does not follow a MASTER SYNC. within a predefined time.

A handshake protocol similar to this is commonly used for the communication protocol needed between computers or between computers and *terminals*; this will be discussed in Chapter 8.

4.4.1.2 *Master–master handshakes*

A similar form of protocol can be used to allow several master elements to share a single set of busses. Supplementary control lines allow the potential masters to compete for use of the busses.

It can be seen from Fig. 4.12 that the master only needs the busses during the actual transaction cycle time. It may not need the busses for some time while it processes or manipulates data. During this free time, some other element of the system can use them. Great care must be taken, of course, to ensure that two or more potential masters never attempt to use the busses at the same time. If other handshake lines are added to the control bus and a suitable protocol is established, the potential masters can share the busses in an orderly manner.

With a two-master bus system, interaction is very simple; a suitable scheme is depicted in Fig. 4.13. Either master can request access to the busses by signalling with its own BUS REQUEST signal but it cannot use them until its own BUS GRANT signal indicates that the other master is 'off' the busses. This very simple handshake protocol allows orderly shared use without any conflicts. Either master can lock out the other (to ensure rapid uninterrupted transactions) by not relinquishing its BUS REQUEST signal.

Several techniques have evolved to allow this simple protocol to be extended to multi-master bus systems. There is usually a single control line (called the BUSBUSY line) which is used by all masters to indicate that the busses are in use. Any master which wishes to gain control over the busses must then request use through one or more BUS REQUEST lines. Some form of *arbitration logic* is required to monitor the BUS REQUEST lines and to decide which requesting master will be given control when the current master is finished. The selected master is given a specific BUS ACKNOWLEDGE (BUSACK) signal from the arbitration logic. When the busses become free, the selected master asserts the BUSBUSY line to become bus-master.

The design of the arbitration logic determines the operation of the system. There are three common structures, the *daisy-chain* (or *priority-chain*), the *polled request* and the *multi-level request* (Fig. 4.14). The daisy-chain structure is the simplest; when the busses are free, the arbitration logic outputs a BUSACK signal to the first master in the chain. If it wants the busses, it 'breaks the BUSACK chain', pulls the BUSBUSY line and uses the busses. If it does not need access it passes the BUSACK signal to the next

Figure 4.13   Handshake between two bus-masters; (*a*) the control signals, (*b*) bus-master transfer signals.

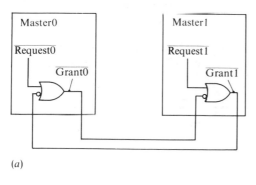

(*a*)

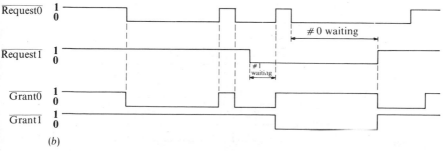

(*b*)

master. The position of a master along the chain determines the priority
that it has for access to the busses.

The fixed priority structure of the daisy-chain bus grant scheme can be
made more flexible by the *polled priority* scheme. This replaces the single
chained grant line with a set of grant lines passed in parallel to all masters.
When the bus becomes free, the bus arbitration logic runs through all
possible patterns on the grant lines in a fixed order; when the grant pattern
corresponds to that programmed into a master requesting access, that
master will grab the buss access and halt the *polling* of the other masters
with lower logical priority. This scheme allows 2^n priority levels to be
generated by n grant lines and allows the priority level of each master to be
manipulated by changing its poll response; but it only allows one master

Figure 4.14   Block diagrams of (*a*) daisy-chain, (*b*) polled request and
(*c*) multi-level request access schemes for multiple masters.

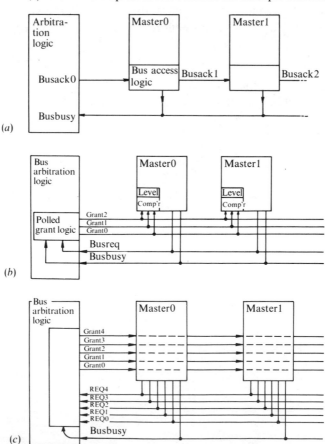

per priority level. The need to cycle through the polling sequence and to wait for a response at each level, makes this technique slow in a long-line bus system.

The polling concept can be improved by another control scheme which uses multiple request and grant lines. In this scheme, the bus arbitration logic compares all the request lines (using fast combinational logic) and issues a grant command on the grant line corresponding to the highest priority request present, as soon as the current master relinquishes the busses.

This scheme is used in the Digital Equipment Corporation (DEC) PDP-11 system and its derivatives. It allows the priority level of any master to be selected by the request and grant lines that it uses; to economise on the number of grant and request lines required, the PDP-11 uses only five request levels but provides the daisy-chain priority scheme at every level.

4.5   Microcomputer bus architecture

This section describes the organisation of the busses, and the allocation of addresses to the various elements, in a typical microcomputer.

The organisation or *architecture* of a microcomputer's bus control system is usually quite simple and has very limited handshake features. In many 8-bit microcomputer systems, the microprocessor is the only master and is either the source or destination of every data transaction.

The microprocessor usually generates most of the control and timing signals of the system from a *system clock* supplied to it by ancillary circuitry. (This is a continuous square wave signal, not a single timing edge.) The microprocessor has its own *internal databus system* which allows data transactions between its internal elements (*registers*), or between these and the external slave elements. The internal and external databusses are connected by a bidirectional multi-bit buffer which allows internal processor transactions to be executed in isolation from the external busses. Fig. 4.15 shows the block diagram of a typical microprocessor.

In its normal role as bus-master, the microprocessor contains at least two registers for driving the address bus. (We will explain the role of all the registers in Chapters 5 and 6.) The R/$\bar{\text{W}}$ control signal output by the microprocessor is intimately associated with the address bus and is buffered in the same manner.

Chapter 5 will describe the detailed operation of the processor; at this stage we will just assert that it must fetch multi-bit words from memory and send modified words back.

The bus system must allow transport of words between the processor and memory or between the processor and the outside world (input/output (I/O) transactions). No distinction has been made so far between memory and I/O transactions for the simple reason that there is none. If the slave element used in a data transaction connects to the outside world, it is an input or output transaction; if the slave simply stores data and returns it on demand, it is a memory element.

The slave element addresses defined by the address bus must be allocated by the computer's hardware designer either to I/O or to memory. The full range of slave elements which can be addressed is called the *address space* of the machine and is allocated or *mapped* as either *memory space* or *I/O space* to suit the intended application of the machine. Two general schemes are used for mapping the I/O addresses onto the address space, *memory-mapped I/O* or *I/O mapped I/O*. The simplest examples of these two schemes are represented by the 6502 microprocessor and the Z80 microprocessor respectively. Box 4.3 describes these two mapping schemes and details the data transaction protocol for the 6502 and Z80.

Figure 4.15   A typical block layout of the bus oriented registers within a microprocessor.

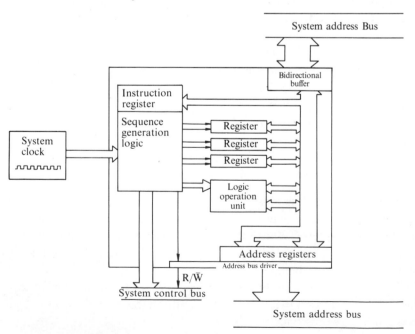

Box 4.3
Microcomputer busses, architectures and data transaction cycles

This box describes the architecture and the data transaction cycles of 6502 and Z80 microcomputers. A detailed knowledge of these features is the first step towards designing control and data collection interfaces.

(1) Memory mapped I/O and the 6502 microprocessor

The microprocessor makes no distinction between the memory address space and the I/O address space. Fig. 4.16 depicts a common representation of a microcomputer's memory space. The memory array is not shown as individual slaves, instead, the full array is usually piled up in one block as if it were a single element with many addresses (Box 4.2). The data, address and control busses make single connections to this pile.

There are two ways, upwards or downwards, to order this pile of binary addresses; both are used, of course. We will see soon that the processor usually executes its program through increasing address locations; we write down a page to avoid blotting our ink, so the software programmer thinks of memory running downwards, numbered from top to bottom. The

Figure 4.16   The 6502 memory layout.

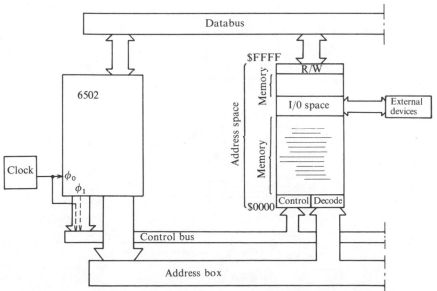

hardware engineer, on the other hand, always thinks of the machine being built from the bottom up. Engineering drawings have a title block in the bottom right corner, where it can be seen in a stack of folded drawings. The list of components on the drawing is placed above this block and must therefore be written upwards (and engineers use pencils!). Hardware layout drawings therefore usually number the memory array addresses from the bottom up.

The timing of the data transaction cycle is determined by the system clock. In a conventional microcomputer, the timing restraints imposed by memory and other requirements impose a lower limit of about 0.3–1.0 μs on the data transaction cycle time. The standard 6502 typically uses a 1 MHz system clock signal input to ϕ_0. It echoes back a propagation delayed clock signal ϕ_1. Fig. 4.17 shows how these two signals relate to the data transaction cycle.

A full data transaction cycle requires one full cycle of the clock. During the first phase of the full cycle (called the *address phase*, ϕ_0 low), the 6502

Figure 4.17   The 6502 data transaction timing showing data movement from and to the processor. Responses to the $\overline{\text{READY}}$ line are also shown.

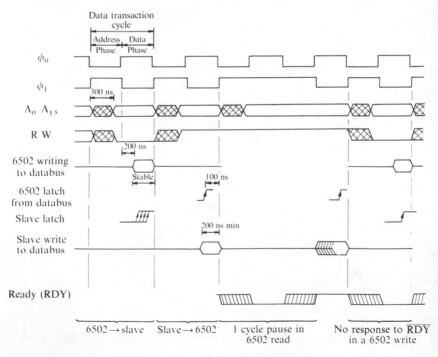

outputs the code of the designated slave on the address bus. The address takes up to 300 ns to become stable. (It will be shown in Chapter 8 that other computer housekeeping functions can be carried out during this phase of the cycle.)

The second phase of the transaction (ϕ_0 high), is called the *data phase*. If the 6502 is writing to the data bus, it will have the data deposited and stable 200 ns into this phase. If it is reading from the databus, it will latch the data off the databus lines about 100 ns before the end of this phase. The 6502 provides a single control line $R/\overline{W}$ to signal whether it intends to read ($R/\overline{W}$ high) or write ($R/\overline{W}$ low) from, or to, the databus.

The 6502 processor does not provide signals to time data deposits or latches to or from the databus. Ancillary timing circuitry must be provided for this within the microcomputer hardware. Typical timing signals are shown on Fig. 4.17. (Relate them to the system depicted in Fig. 4.10.) If the timing signal propagation delays in the 6502 and the slave logic are considered carefully, it is possible to use the falling edge of the ϕ_0 clock to latch data off the bus into the slave.

A data transaction cycle in the 6502 cannot be performed in less than the clock period. To allow operation with slave elements which are unable to respond within this time, a simple handshake control line called $\overline{READY}$ is provided. In normal operation, it is left at logic high. If it is pulled low during the address phase of a 6502 READ cycle, the processor will not enter the data phase to carry out its data transaction until the control line is restored to logic high. The data transaction will then be completed during the next ϕ_0 clock high period.

The logic level of the $\overline{READY}$ line should only be changed while ϕ_0 is low or at the start of the data phase. The $\overline{READY}$ control line has no effect during a 6502 WRITE transaction cycle.

(2) I/O mapped I/O and the Z80 microprocessor

The Z80 microprocessor allows two separate memory spaces to be accessed by its 16-bit address bus. The primary space is called the *memory space* and uses all 16 bits; this allows 64K of memory elements to be designated. The secondary space is called *I/O space* and uses only the lower eight bits of the address bus (A_{0-7}); this space therefore defines only 256 separate I/O addresses. The microprocessor selects from one or the other of these spaces by using one of a pair of signal lines in the control bus. A conventional Z80 microcomputer's memory layout is shown in Fig. 4.18.

The Z80 provides fully timed signals for reading and writing from and to the databus (Fig. 4.19).

The Z80 does not give a specific signal indicating whether it will read or write to the databus (cf. the 6502), instead it delivers a fully timed signal on either its $\overline{\text{READ}}$ or its $\overline{\text{WRITE}}$ control line. The two control outputs, $\overline{\text{IOREQ}}$ and $\overline{\text{MEMREQ}}$, are used to signal whether the microprocessor expects to communicate with the I/O or the MEMORY space respectively.

The data transaction timing of the Z80 is more elaborate than that of the 6502. The system clock frequency is about 4 MHz for the standard Z80; the transaction cycle is divided into several sequential phases corresponding to cycles of this clock, called T_1, T_2, T_3, etc.

Fig. 4.19(a) shows the control signal sequence for a Z80 read and write from and to memory space. Note that the control signals provide complete timing and control. The data transfer is carried out during T_3 to allow two full clock cycles for the busses to become stable and for the slave element to respond.

During a transaction with the I/O space the timing is similar to that for memory, but the Z80 only uses the lower eight address bus lines; it signals to the I/O slave elements by pulling the $\overline{\text{IOREQ}}$ line low instead of the $\overline{\text{MEMREQ}}$ line and it adds a further phase, T_W (a WAIT phase), between the T_2 and T_3 phases to allow the I/O elements an additional 250 ns response time.

Figure 4.18   A typical block diagram of a Z80 microcomputer.

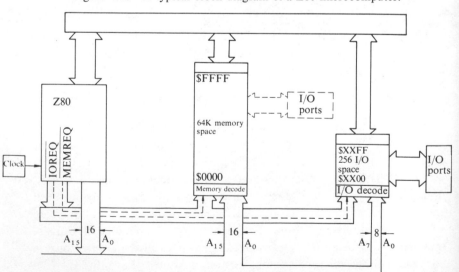

The provision in the Z80 of a separate I/O space, does not prevent I/O elements from being mapped into memory space provided that device timing is considered.

Fig. 4.19(c) shows the simple handshake that allows a slow slave to delay a data transaction. The Z80 consults the $\overline{\text{WAIT}}$ line on the falling edge of the system clock during T_2. If any slave element is holding this line low at that instant, the Z80 stalls the transaction by carrying out dummy cycles (T_W phases), consulting the $\overline{\text{WAIT}}$ line on each falling clock edge until the

Figure 4.19 The Z80 data transaction protocol: (a) a memory transaction, (b) an I/O transaction and (c) handling slow access with WAIT states.

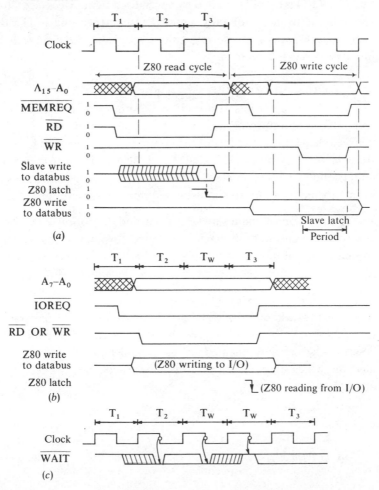

WAIT line goes high again. The Z80 then enters the T_3 phase on the next rising clock edge.

The provision of the second I/O address space in the Z80 architecture shows its different evolutionary path from that of the 6502. The Z80 microprocessor evolved from the first microprocessors (the 4004 first made by INTEL in 1971); these were extensively used in engineering and control applications, where there are a lot of data-input/control-output transactions between the microprocessor and outside-world systems. This background explains the Z80 address architecture and its fully timed I/O signals. The 6502, on the other hand, was developed by Synertek as a microprocessor for use in a small microcomputer. In such an application, the role of I/O is usually handled by specialised I/O devices (see Chapter 8) and there is no need for fully timed I/O control signals. The different evolutionary paths of these two microprocessors are also seen in their instruction sets (Chapter 6).

Closing remarks

This chapter has explored how the *bus structure* of conventional computer architecture allows multi-bit binary words to be transported between the various elements of the computer. The protocol for these *data transactions* has been explained in sufficient depth to give a basic understanding of the operation of bus systems ranging from very simple 8-bit microcomputers up to sophisticated mini- and mainframe computers.

The concept of master and slave elements, and the notion of transfer of bus control between multiple masters has been explained. These aspects of computer operation will be explored further in Chapters 7, 8 and 9, where we will examine some powerful general techniques for connecting a computer to the outside world for data-collection and control operations.

In the next two chapters we will delve deeper into the internal operation of the microprocessor itself, to see how the data transaction hardware described above is augmented by additional logic circuitry, to create a machine which can execute programmed, predefined sequences of logic operations.

Exercises

4.1 Describe the architecture of the bus system used in a computer for multi-bit word transactions.

Using the simplified diagrammatic model of Fig. 4.10, describe the procedure used for data transactions; show simplified timing diagrams to indicate the signals issued by the bus-master while it controls the transaction process.

4.2 Discuss the advantages and disadvantages of the bus system of Exercise 4.1 for a computer. Why is the bus transaction scheme used universally in computers?

4.3 Explain why there must be a master and a slave during every transaction in a bus system. Discuss how multiple masters can be accommodated in such a system.

Describe some schemes which allow several devices to become masters and to share use of the busses.

4.4 Consider the extension of the master–slave relationship discussed in this chapter in terms of internal transactions, to the general problem of data transactions (communication) between generalised data systems (Chapter 8).

References and further reading

1 *Mini/Microcomputer Hardware Design*, G. D. Kraft & W. N. Toy, Prentice-Hall, New Jersey (1979)
2 *Z-80 User's Manual*, J. J. Carr, Reston-Prentice-Hall, Reston, VA (1982)
3 6502, Z80, 6809 etc. data and applications manuals (Synertek/ Rockwell, Zilog and Motorola)
4 Motorola etc. ECL data manuals

5

The instruction execution cycle

Preamble

Chapter 4 described how simple logic circuitry within the processor allows it, as bus-master, to control data transactions across the bus system of the computer. This chapter will explain how further straightforward (but extensive) logic circuitry within the processor allows it to merge multi-bit data transactions with simple logic operations. It will also show how the careful design of this ability endows it with remarkable versatility. It will show how the internal architecture of a microprocessor achieves this, using only the simple logic operations described in previous chapters. We will explain how the processor fetches code words from memory and uses them to determine the operations that it is to perform. Program execution will then be seen to be what it really is, an automatic, mindless machine response to simple instructions arranged in memory as a string of code words. Chapter 6 will describe and explain the limited set of sequences (instructions) that a microprocessor can execute, and will show how this limited set is able to provide almost unlimited versatility in the machine.

During the development of the courses on which this book is based, it was found that the several new concepts involved in this aspect of computer operation could be best understood if they were explained first for a single microprocessor; it was also found that the 6502 microprocessor was the easiest to comprehend. These two chapters will therefore be restricted to a discussion of the 6502; this chapter will concentrate on its instruction cycle sequence and the next chapter will concentrate on its set of instructions. Reference will be made to the Z80 (and others) when necessary, to point out some general feature which is not obvious or present within the 6502. Once the contents of these two chapters are mastered, the reader will find it very easy to transfer the concepts to any other microprocessor, using only the manufacturer's data manuals.

5.1   The microprocessor's internal structure

A brief outline of a microprocessor's architecture was given in Section 4.5. We will now explore the 6502 internal architecture in sufficient detail to understand its operation as a binary logic machine. At first sight it looks a bit intimidating (Fig. 5.1), but when the purpose of each of the many elements is explained, its overall operation will be seen to be quite simple and obvious. Note first that the internal 8-bit databus allows data transactions between the internal registers, or between one of these and the external databus.

The *Instruction Decode Logic* (*IDL*) block is the control unit of the microprocessor. It operates the external control bus of the computer and thus is the control unit (the master) of the entire computer. The IDL block has *direct*, discrete control connections to each internal register; this direct control avoids the need for an internal address bus to access and command

Figure 5.1   The 6502 internal architecture.

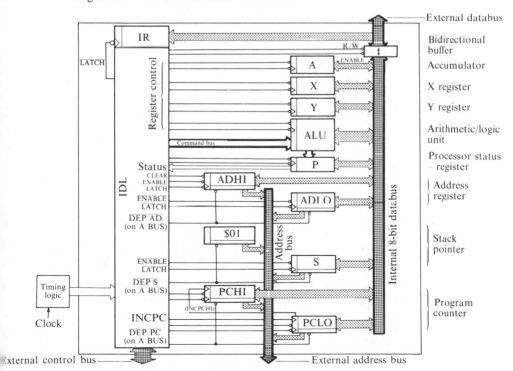

the internal registers and allows data transactions between the internal registers to be carried out very quickly.

Three pairs of 8-bit latches or registers are provided with driver gates for the 16-bit external address bus; these are the Address Register, the Stack Pointer and the Program Counter. Full 16-bit addresses can be assembled in these registers by transactions over the internal databus; the IDL can force one of these addresses onto the external address bus to nominate an external slave element for a data transaction. This arrangement, combined with discrete control lines to each internal register, allows the IDL to carry out data transactions between internal elements or between any external element and an internal element. The execution of a data transaction cycle on a bus system by a timed control signal sequence was described fully in Chapter 4. The adaptation of the sequence to use a specific internal element is obvious.

The IDL can execute many different control sequences, each composed of data transaction cycles and simple multi-bit logic operations. These sequences are called *instructions*. The sequence to be executed is determined by the word present in the *Instruction Register* (IR).

   An instruction always begins with a data transaction to fetch a word into the IR. The IDL then carries out the control sequence defined by that word. When this is finished, a new word is fetched into the IR. Thus it automatically executes a set of predefined control sequences defined by a sequence of words in memory. In computer *software* terminology, the microprocessor can execute a program of instructions held in memory. The instructions that will be discussed in this chapter are those that are used directly by the microprocessor within its IDL; they are said to be in *machine language*.

   In the following sections of this chapter, a few simple control sequences (instructions) involving some of the internal registers will be described. The full range of instruction sequences and the significance and purpose of the other registers will be described in Chapter 6.

5.2   The instruction execution cycle

   Instruction execution is simply the machine response to a sequence of control signals issued by the IDL block. We will describe first the signal sequences needed for execution of some simple instructions, and we will then delve deep into how the IDL performs its magic.

When the 6502 IDL starts an instruction cycle, it always first issues a signal sequence to fetch a code word into the IR. This is performed by the IDL forcing the address in the *Program Counter Register* (*PC*) onto the address bus, and then issuing the control sequence to execute a data transaction cycle from the addressed memory into the IR. When this is finished, the IDL issues a control signal to increment the address held in the PC.

Let us assume that the PC contains $7000 (0111000000000000), and that this location in address space is a memory element which contains the word $A9 (10101001). As a result of this first data transaction of the instruction execution, the IR will hold the word $A9. The presence of this word causes the IDL to perform a further sequence of predefined control signals. The full sequence, including the common first steps, is as follows.

   (1) The word at the memory location specified by the PC is moved to the IR (common first step).

   (2) The PC is incremented by one (common first step).

   (3) A second data transaction is carried out using the address specified by the PC as the source and the A register (Accumulator) as destination.

   (4) The PC is incremented again.

This sequence is illustrated schematically in Fig. 5.2; note that the address space is shown in the software model. This simple control sequence copies the word in memory element $7001 into the A Register. After the sequence

Figure 5.2   Execution of the sequence defined by $A9: (*a*) simple model of the computer; (*b*)–(*e*) Steps (1)–(4).

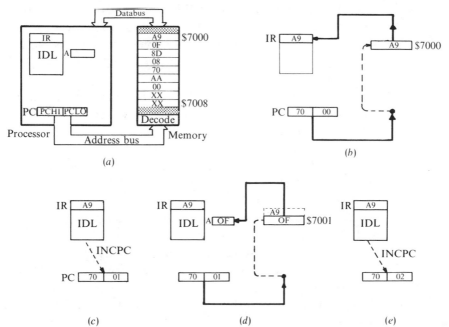

is finished, the PC is left pointing at the next memory element. The IDL logic moves immediately into another instruction cycle by fetching the data at that element into the IR and executing the sequence defined by that word.

Before moving on, let us review the control signals that the IDL unit must generate for the simple instruction execution described above. At Steps (1) and (2), the IDL must carry out the following sequence:

command the PC to deposit its contents on the address bus,

command the designated slave element to deposit data,

command the bidirectional buffer to read in,

command the IR to latch data from the internal databus and, finally,

increment the PC contents.

At Steps (3) and (4) the IDL must carry out a command sequence almost identical to that for Steps (1) and (2), but this time it commands the Accumulator to latch data off the internal databus. All of the control lines needed can be traced on Fig. 5.1 from the IDL to the registers; later in this chapter we will describe some simple circuitry to generate them within the IDL.

Let us now return to where we left the microprocessor, with its PC '*pointing*' at address $7002. Let us assume that this address holds the data $8D. When the IDL enters the next instruction cycle, it carries out the more elaborate sequence defined by that word and described below (see Fig. 5.3).

(1) Move the contents of the memory location specified by the PC to the IR.

(2) Increment the PC.

(3) Execute a data transaction cycle using the addressed memory element ($7003) as source, but the *ADdress LOwer-value byte* (*ADLO*) register as destination.

(4) Increment the PC.

(5) Execute a data transaction cycle using the addressed memory element ($7004) as the source and the *ADdress HIgher-value byte* (*ADHI*) register as the destination.

(6) Disconnect the PC from the address bus, and connect the ADHI, ADLO register pair to the address bus.

(7) Execute a data transaction cycle using the A Register as source and the memory element addressed by the *address latch* (ADHI, ADLO) as destination.

(8) Disconnect the ADHI, ADLO pair and reconnect the PC.

(9) Increment the PC.

This sequence will copy the two words in the memory elements $7003, $7004 into the ADLO, ADHI register, assembling them into a 16-bit address; the IDL will then copy the contents of the A Register into this new address and will return to the memory element $7005 to begin the next instruction cycle.

Let us assume that the element $7005 holds data $AA. When the IDL continues into the next instruction cycle, the following sequence is carried out.

(1) The word at $7005 is moved into the IR.
(2) A data transaction is carried out within the microprocessor, over the internal databus; the A Register is the source and the X Register is the destination.

Figure 5.3   Execution of the sequence defined by $8D: (*a*) the simple model; (*b*) Step (1); (*c*) Steps (2) and (3); (*d*) Steps (4) and (5); (*e*) Steps (6) and (7); and (*f*) Steps (8) and (9).

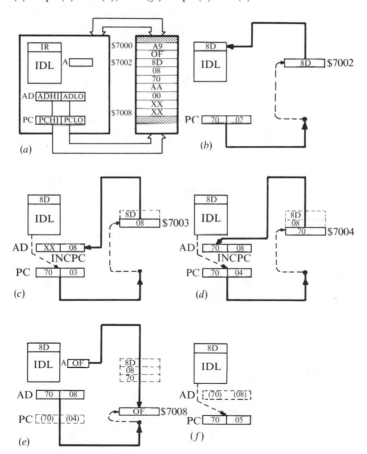

(3) The PC is incremented by 1.

This simple control sequence is depicted in Fig. 5.4; it causes the contents of the A Register to be transferred (copied) into the X Register. The PC is left pointing at the next memory location ($7006).

The word addressed by the PC will again be loaded into the IR in the first step of the next instruction cycle. This word is shown to hold $00 in our examples. The IDL requires no further memory transactions to execute the sequence defined by this word. It causes the IDL to cease execution of this little program segment and to go elsewhere in memory to collect the next instruction. (The manner in which this is done will be explained in Chapter 7.)

Before we delve into the inner works of the IDL, we should note a few generalisations about the sequences just described.

> The first two instruction sequences consisted of a *codeword* which defined the sequence to be carried out, followed by one and two words, respectively, used in the execution of the sequence. The latter two instruction sequences required only their instruction code word.

> The code word specified the precise control sequence to be carried out, and therefore had implicit within it the number of words that were needed when the coded sequence was executed.

Figure 5.4   Execution of the sequence defined by $AA: (*a*) the simple model; (*b*) Step (1); (*c*) Step (2); (*d*) Step (3).

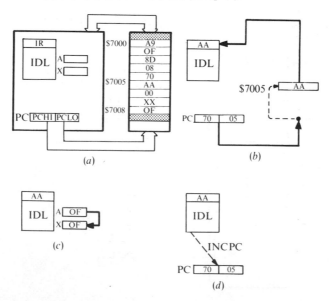

The code word for the subsequent instruction must follow directly in memory after the last word of the previous instruction.

A 6502 microprocessor instruction is thus a group of one, two or three words placed sequentially in memory. The first word of the group defines the operation to be carried out. It is called the *operation code* (or *opcode*) of the instruction. The opcode implicitly defines the number of data words that must follow. These extra words are called the *operand* of the instruction. In the examples given so far, they define the source or destination of the data which is to be transferred to or from the A Register of the microprocessor. (We will see further uses of the operand in Chapter 6.) The instructions follow each other in ascending order of the address space elements. This is the essence of the *von Neumann architecture*; the machine executes one instruction at a time in a sequence defined by commands and their arguments, organised as a string or *program* in memory.

The manner in which the commands use their arguments can be very flexible; consider the three command codes $A9, $8D and $AA. With the code $A9, the *data* is at the address that immediately follows the opcode; with the code $8D, the destination *address* follows the opcode in the next two words (low byte first, high byte second); the code $AA has the source and destination defined implicitly by the opcode. This aspect of instruction execution is called the *addressing mode*. We will discuss it in depth in Chapter 6. The first mode is called *immediate addressing*, the second is called *absolute addressing* and the third is called *implied addressing*.

The IDL can only understand instructions in its own language, that is, as the code words used to define the instructions. These programs of instructions, consisting of a string of such code words (opcodes) and their associated data words (operands) are called *machine code* or a machine language program. A string of such binary words or, more conveniently, their hexadecimal equivalents, are rather hard to interpret or remember. To make the machine code more recognisable and memorable, each instruction opcode is assigned a *mnemonic* representation and each addressing mode is assigned a symbolic representation. Box 5.1 gives a brief overview of these aspects; they will be fully described for the 6502 in Chapter 6.

Box 5.1
Machine language and mnemonic code————

The pure binary machine code, the hexadecimal equivalent and the mnemonic code for the short program described in the main text are shown in this box. The group of binary words comprising each instruction is given as one line of hexadecimal code and a single mnemonic representation. Only three of the many possible addressing modes are shown here; the other modes and their mnemonic representation will be given in Chapter 6.

Memory address	Binary machine code	Hex. code	Mnemonic representation
$7000– $7001–	10101001 00001111	A9 0F	LDA #$0F
$7002– $7003– $7004–	10001101 00001000 01110000	8D 08 70	STA $7008
$7005–	10101010	AA	TAX
$7006–	00000000	00	BRK

The mnemonic code is very easy to understand and remember. The program representation on the far left can be easily read:

LoaD the Accumulator (LDA) with the data $0F using *immediate mode* (the data immediately follows the opcode).

STore the contents of the Accumulator (STA) at the address specified in *absolute mode* by the following two words.

Transfer the contents of the A Register to the X Register (TAX).

And then BReaK execution (BRK) of this program.

This mnemonic code is obviously very convenient to describe the operation of the program sequence. High-level Editor/Assembler programs which allow machine language code to be written in this form are freely available for every microcomputer. These allow the writing and editing of the mnemonic code as plain text using the text editor. The resultant text code can then be used as the input source for the assembler section of the high-level program; this converts the text of the *source code* into the machine language code (*object code*) that the microprocessor can execute.

In this chapter and the next, machine language routines will be restricted to those that are simple enough to be coded directly into machine language 'by hand'. There are two prime reasons for this restriction. Firstly, it avoids the need to specify and learn the syntax and rules of a high-level language (the Editor/Assembler commands); secondly, the purpose of this book is to give a sound understanding of the machine's operation at the hardware level. The reader will find that direct machine language coding with a simple microprocessor like the 6502 gives a thorough understanding of the microprocessor's internal operation; Editor/Assembler programs then become eminently understandable and full use can subsequently be made of their powerful facilities.

5.3 The IDL

This section describes simple circuitry within the IDL to generate the sequence of control signals needed by the 6502 microprocessor during execution of two of its 152 instructions (two of those described in the previous section). The circuits described are intended only as a simple model to demonstrate that the hardware of a computer has no intelligence at all; it is nothing more than a rather complicated assembly of simple, mindless logic circuits.

It is obvious that great care must be taken in the definition of the control sequences so that the microprocessor is able to perform efficiently the greatest possible range of programs. (This will be considered in Chapter 6.) The designer must cram as much as possible into the IDL; thus the circuits used there must be optimised not only for efficient use of the limited space, but also to facilitate debugging and development. The general approach used for a real processor will be outlined in Chapter 8.

The IDL generates a unique, predefined sequence of control signals in response to the presence of each individual opcode in the IR. The control signal sequences are called the *instruction interpretation* of the IDL; it can be thought of as a set of procedure cards, one of which is selected by the opcode in the IR, the procedure listed on that card then being carried out by the IDL.

A description of the *effect* of all of these sequences or procedures is called the *instruction set* of the processor.

5.3.1 A simple model of the 6502 IDL

We will explore a small part of the logic within the IDL block shown as a black box on the left of Fig. 5.1.

The instruction interpretation sequences generated by the IDL are timed by signals derived from the timing logic attached to the IDL. Fig. 5.5 shows the output signals required (the diagrams are simplified for clarity). The timing is similar in principle to that described in Chapter 4 (Fig. 4.11(*a*)) and in Chapter 2 (Fig. 2.20); it uses a set of cycle counter signals to control the sequence of operations.

The IR of the IDL is connected to a decode logic block similar to that discussed in Chapter 1. Every allowed opcode (bit pattern) in the IR, causes one specific output line to go high. Fig. 5.6 shows how this decode circuit selects one of the many predefined sequences to be selected for execution. The logic blocks of Fig. 5.6 are arranged in columns corresponding to the different opcodes and rows corresponding to the logic signals generated during each successive cycle of the instruction execution.

For the sake of simplicity, Fig. 5.6 is shown with a lot of redundancy. For clarity, the control output gates of the IDL are shown as open-collector logic. The circuit shown *models* the behaviour of the real 6502; it is not meant to represent it accurately. Note that the first action in the execution of an instruction is always to fetch the word from the element addressed by the PC into the IR. (See the note at the end of this section.) Also note that at the end of each instruction interpretation, the timing logic is reset to the first cycle, thus forcing immediate execution of the next sequence. The IDL will therefore run continuously fetching opcodes and interpreting them. If the IDL loads itself with an opcode for which no sequence is defined, it will 'crash' and will be unable to resume operation. This could happen when the machine is first turned on, or if the program were not properly written or constructed. Provision must be made for an orderly start-up when the machine is first turned on, and for recovery from crashes; we will explore both of these in Chapter 7.

Fig. 5.7 displays the timing diagram for the execution of the $A9 sequence. The internal operation of the IDL can now be seen to be a simple elaboration and extension of the data transaction cycle described in Chapter 4.

Note: in both the 6502 and the Z80, an output signal is sent to the control bus when the processor is fetching a new opcode into the IR. In the 6502, it is called SYNC; it is held high by the 6502 during the opcode fetch transaction cycle. In the Z80, the signal is called M1; it is held low during T_1 and T_2 of the instruction opcode fetch.

Figure 5.5   The timing and control signals required for the IDL of Fig. 5.6. See also Figs. 4.11(a) and 2.20.

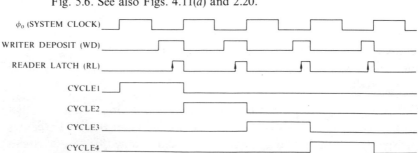

Figure 5.6   A simplified model of the IDL circuitry to generate control sequences for the codes $AA and $A9.

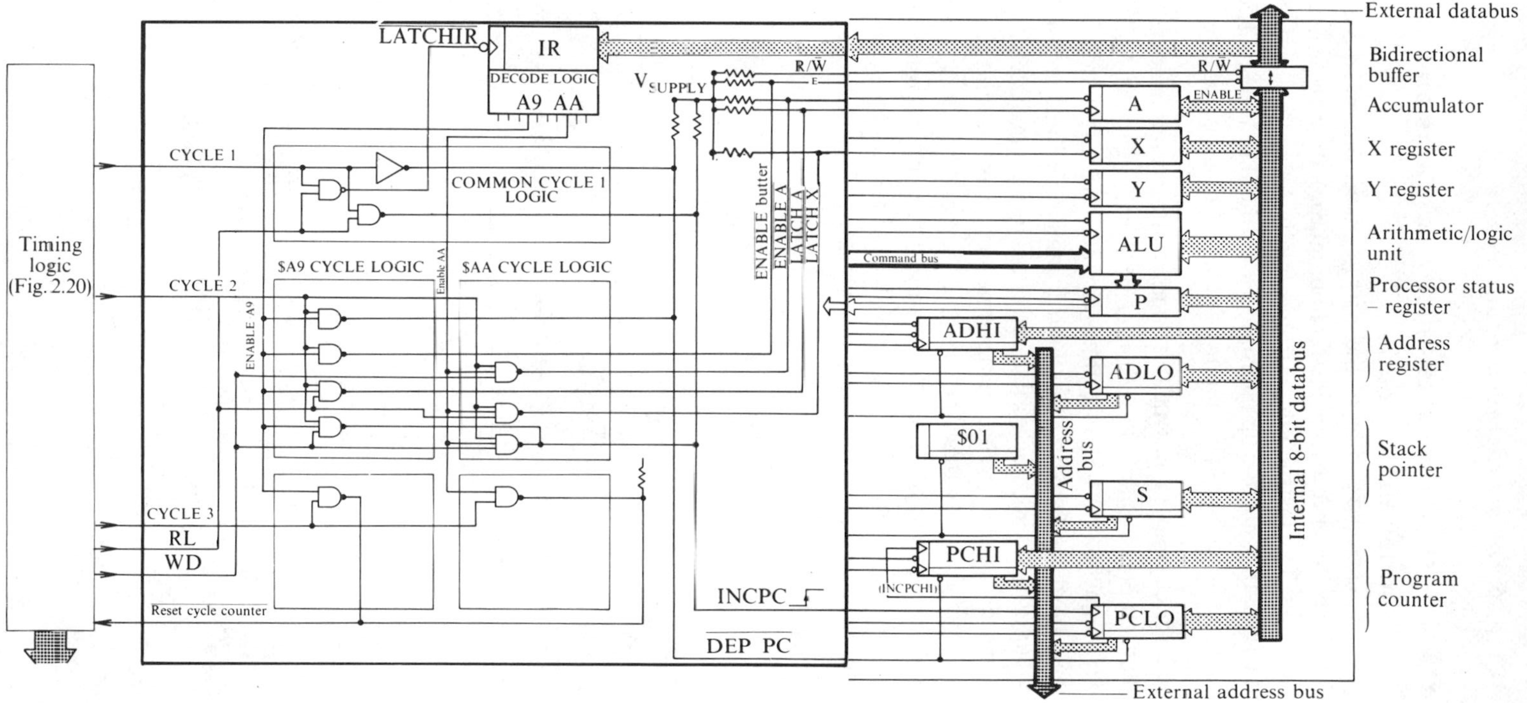

5.4   Instruction sets, microprograms and machine independent code

Section 5.3 showed the instruction set of a microcomputer to be completely defined by the instruction interpretation sequences that it can carry out. The instruction set is thus totally defined within the circuitry of the IDL. In a single-chip microprocessor, the IDL is contained within the chip's microcircuit and cannot be accessed or modified from outside.

In large computers, instruction interpretation is implemented in a part of the computer called the *Central Processor Unit* (*CPU*). Before the development of VLSI integrated circuits, the instruction interpretation logic block of the CPU was built from simple logic gates. This type of circuitry was readily accessible and could be designed such that it could be changed quickly. The instruction set was usually built into an interchangeable part of the machine. Several 'prewired' instruction sets could be provided. For the manipulation of a large database, the instruction set could be tailored with emphasis on fast access to memory and quick comparison of multiple words; for scientific or mathematical

Figure 5.7   The control sequence timing for $A9 instruction code.

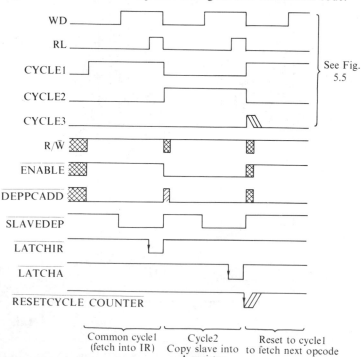

work, the emphasis could be on operations tailored to suit rapid arithmetic.

As the machines became more sophisticated, a concept called *microprogramming* evolved. This replaced the large instruction set of the previous CPU with a smaller set of very simple instructions held within the permanent instruction decode and interpretation logic. A fast, cache memory module was added to the CPU; this was used to define short sequences of the permanent instructions and thus to define a set of more complex, compound instructions (microprograms). The micropro-grammed instruction set, held in fast memory within the CPU, could be loaded or modified by software; the user could thus define his own instruction set! A major advantage of this was that the instruction set of another processor could be *emulated* by a microprogram and all software written for the other processor could then be run on the microprogrammed machine.

The microprogram concept represents a two-tiered instruction set. The general principle is frequently used nowadays to provide *machine independent software*. The high-level language PASCAL is structured in this form. Its 'machine' language program is expressed in a set of instructions (opcodes and operands) defined for a machine called the *P-machine* or *Pseudo-machine*. There is no such real, hardware machine. PASCAL is implemented in any computer by a small set of program segments written in the host's machine language, to emulate the instruction set of the P-machine. This allows the host to interpret P-code programs, and allows the entire structure of the high-level language to be designed and perfected once only. Any PASCAL program is fully *transportable* to any computer. The logical separation of the P-code from the host's hardware can, however, cause some problems if the user wants to access the bus directly for data collection and control operations.

5.5 Some general remarks on processor philosophy

It is obvious that program execution requires every word to be absolutely correct and in the correct place; it can be likened to moving along a narrow parapet above a bottomless abyss. The machines that we have discussed so far have no facility for recognising, correcting or recovering from errors; they require absolute precision in their instructions. The importance of the noise rejection inherent in digital logic (Chapter 1) to the proper functioning of a computer should now be obvious. Under normal conditions, the *hardware* of the machine will always function perfectly, that is, without error. The intellectual input, the program, must also be perfect (without a single error) for the hardware/software system to work perfectly.

Section 5.3 showed common microprocessors to have their instruction sets buried within internal logic circuitry; they cannot be changed. The designers must therefore take great care in the definition of the microprocessor's instruction set; it must represent the best possible compromise for all of its many, and often conflicting, applications.

You will recall from Chapter 1 that virtually any combinational logic expression can be created from a few very simple logic operations. In the next chapter, we will explore the instruction set of the 6502 microprocessor and we will find that, analogous to combinational logic, virtually any task that can be represented as a sequence of logical operations can be performed by a machine that can execute a small set of carefully chosen but simple logic operations on multi-bit words. This transforms the simple-minded machine described in this chapter into an extremely versatile and powerful tool for use in an enormous range of applications. In subsequent chapters we will see how this power and versatility can be developed to a high level of sophistication and can be extended to interactions with the outside world.

When writing programs for a microprocessor, it is seldom necessary to understand fully its internal circuitry. From here on we will generally ignore the internal operation of the IDL and will concentrate on the microprocessor's functional characteristics by regarding it as a black box. However, the explanation given in this chapter will allow the reader to comprehend the microprocessor as a simple, eminently understandable machine and will allow the content of subsequent chapters to be properly comprehended.

At this stage we have explored the basis of all the wisdom and intelligence of fourth-generation computers! The rest of this book needs only to add some details and to describe some enhancements and refinements.

Mankind has nothing to fear from computers themselves; they are totally impotent and mindless, with no innate intelligence at all. As such, they demonstrate the wisdom and intellect of homo sapiens ... Let me hasten to add that they demonstrate this in the subtlety of their conception and design, in their ability to be commanded (by real, human intelligence) to perform an endless variety of tasks. Any fear of the power and influence of computers must be directed not at the machines themselves, but towards those who control them, and who determine their applications and programs.

Exercises

5.1 Describe the control sequence generated by the IDL to carry out the instruction sequences shown in Figs. 5.3 and 5.4.

5.2 Draw a timing diagram like that of Fig. 5.7 to show the control sequence generated during the execution of the instruction code $AA. Do the same for the instruction code $8D.

5.3 Attempt to devise a simple logic timing circuit similar to those shown in Fig. 5.6 to generate the control sequence for the instruction code $8D.

5.4 Consider the groupings that could minimise the logic circuitry of the IDL of Fig. 5.6.

5.5 The *absolute* addressing mode used by the instruction code $8D is also used by 14 other codes in the 6502. Attempt to devise a scheme which allows a single block of logic circuitry, shared by all of these codes, to establish the slave address, but which allows the IDL to define the direction and the internal register to be used in the transaction.

References and further reading

1 *SY6500 Microprocessor Family – Applications Information* AN2, Synertek (USA), 1982
2 *Computer Architecture and Organization* (ibid. Chapter 1)
3 *Mini/Microcomputer Hardware Design* (ibid. Chapter 4)
4 Z80 hardware reference manuals, e.g. the Osborne series on microprocessors.

6

The instruction set of a microprocessor – machine language programming

Preamble

Chapter 5 described computer operation at the logic signal level to show the basic simplicity of instruction execution. In this chapter, we will ignore the electronic logic; we will consider only the outcome of the IDL control signal sequences. This approach is called the *programming model* of processor operation. We will explore the full set of instructions provided within a simple microprocessor to see how a simple machine, with a small but well-designed repertoire of inbuilt instructions, can be very versatile. As in Chapter 5, and for the same reasons, we will discuss only the 6502 microprocessor in detail. Aspects that are not adequately demonstrated in the 6502 will be discussed with reference to other microprocessors or computers.

Chapters 1 and 2 described how virtually any logic operation could be created from a small set of simple elementary logic operations. A *complete set* of elementary operations must be able to generate any logic operation expressible in the formalism. We saw that the practical **NAND** or **NOR** logic gate (with an inherent propagation delay) could provide all combinational and sequential digital logic functions, and is sufficient to create a computer's hardware. A similar notion of completeness can also be applied to logic operations on multi-bit words; a set of elementary multi-bit operations can be postulated from which all higher-level operations can be constructed. A machine provided with these basic operations would be able to perform any logic task and thus would be able to solve any logic problem for which a solution existed. The only limitation to such a machine would be the time and memory space needed to complete the task. Turing, in the 1930s, developed this hypothesis in relation to an imaginary *Turing Machine* to show that there are many logical and mathematical problems which have no solution.

This chapter develops the notion of a computer's instruction set as a complete set of elementary operations. It is presented as a carefully structured introduction to machine language programming. The topic is usually considered to be very difficult, but the reader should find that the mechanistic explanation presented in previous chapters makes it all seem quite obvious and easy to comprehend. There is, nonetheless, quite a lot to absorb; the range of elementary operations required to construct a complete instruction set is considerably more extensive than that for the binary logic explored in Chapters 1 and 2. Furthermore, the structure of machine language contains several concepts, all of which must be grasped before the topic can be properly understood. This chapter presents these in the order that has been found most effective in undergraduate courses. Note well, however, that, even when all of these aspects are fully understood and the complete instruction set is known, the machine language programmer must still learn how to translate a task into a sequential logic algorithm and to express that algorithm using only the minimal set of operations provided by the machine. Efficient machine language programming therefore requires a fluent knowledge of the available instruction set *and a lot of experience*, particularly in techniques for testing and debugging software.

This knowledge and experience is quite portable from microprocessor to microprocessor; the reader will find that once the 6502 has been mastered, the techniques and expertise developed on it can be transferred easily to other microprocessors and, indeed, even to large computers. The 6502 has a very simple instruction set, yet it displays most features of more complex machines in a rudimentary and easily understood manner. It therefore represents an excellent little machine on which to first master machine language programming.

In this chapter we will work directly in machine code. This is only possible on a microcomputer with a very simple instruction set and, even there, it is restricted to short programs or routines. For many microprocessors (especially 16-bit machines), and for long programs, it is essential to create the machine language program code with an Editor/Assembler high-level language program. (This will be discussed in Section 6.7.) In this book, where machine language programming is only a small part of the content and scope, we therefore restrict examples and exercises to those which can be coded directly into machine language. After working through this chapter, the reader will find it easy to understand and use an Editor/Assembler program.

We will start by describing the registers within the 6502 microprocessor and explaining their roles in the execution of instructions. The subtle use of these registers to create a range of addressing modes will next be described. After this, we will work through the instruction set, giving examples of how each is used.

6.1   The processor registers

The IR and the IDL were described in detail in Chapter 5 to explain the 'machinery' of instruction execution. We will now ignore this and consider only the *effects* of instruction execution. This *programming model* describes the machine as seen from 'outside'. Fig. 6.1 depicts the program model of the internal registers of the 6502 microprocessor; compare it with Fig. 5.1.

The 6502 microprocessor has three 8-bit registers which are intimately involved in data transactions and logic operations during program execution. These are the *A Register* (the Accumulator), the *X Register* and the *Y Register*. We have seen in Chapter 5 that data transactions from or to memory elements are always made between the memory (slave) element and the microprocessor (master). These three registers can be used for these transactions; there are many instructions provided for moving data between them and memory. They are called *general purpose registers*, but

Figure 6.1   The program model of the 6502 microprocessor.

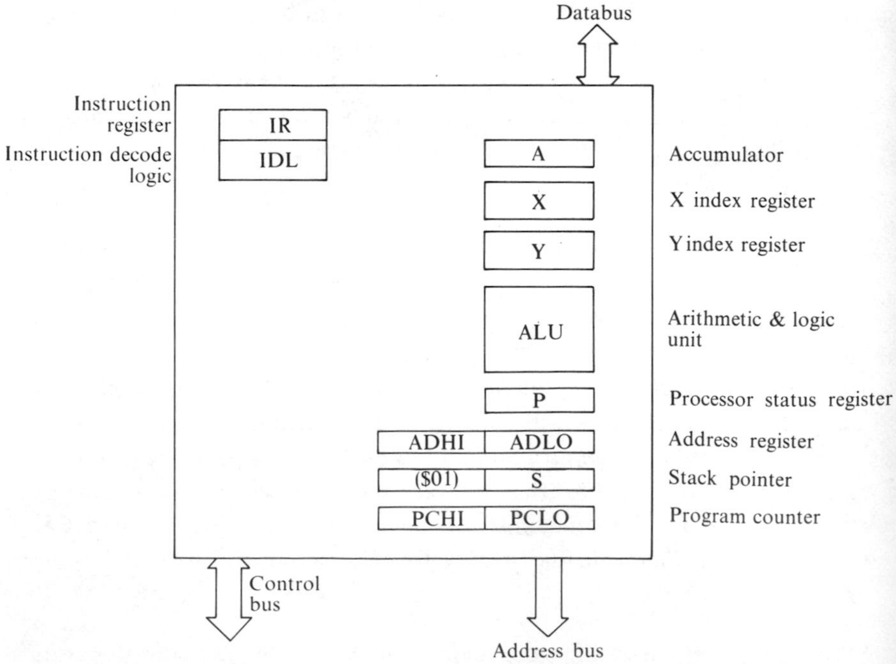

they all are somewhat specialised. The Accumulator is preferred for logical operations; there are many instructions provided in the instruction set for performing logical and pseudo-arithmetic operations on data within it. The X and Y Registers are intended principally as *index registers*; we will explore this in the next section.

The 6502 contains three 16-bit registers which can drive the address bus to designate a selected address for a data transaction. These are the PC, the Address Register and the Stack Pointer (S Register). The PC is normally used by the IDL to run the microprocessor through the program being executed.

The Address Register is used, as described in Chapter 5, as a supplementary register during instruction execution to point to the source or destination of data for a data transaction. The S Register is actually a 16-bit register, but the top 8 bits always contain **00000001** ($01); it can therefore only designate addresses between $0100 and $01FF, PAGE # 1 of the address space. The specialised role of this address register will be described in the next section.

The *Arithmetic and Logic Unit* (*ALU*) cannot be directly accessed by any instruction; the IDL uses this unit to perform multi-bit logic operations on data words passed to it as *arguments* during execution of logical and arithmetic instructions.

The 8-bit *Processor Status Word* (*P Register*) is connected to the internal databus and to the ALU. The individual bits of this register are usually called *status flags* or *status bits*; some of these indicate specific results of a data transaction or logic operation within the microprocessor and its ALU. The execution of each instruction of the set affects this register in a different manner. Some instructions have no effect at all, but most of the instructions effect a predefined subset of the bits. Table 6.1 (p. 182) summarises the entire behaviour. We will describe the full role of this register and will explain the use of Table 6.1 when we are working through the full instruction set later in this chapter; at this stage we will only describe the significance of each flag or bit (Fig. 6.2).

D_7, N, the m.s.b., is the Negative Result Flag. This bit indicates, if effected by an instruction's execution, whether the most significant bit of the resulting data was true (a negative number result ($D_7 = 1$) or false (a positive result) ($D_7 = 0$).

D_6, V, the Overflow flag indicates (if effected) whether a two's-complement signed integer arithmetic overflow occurred ($D_6 = 1$) or not ($D_6 = 0$).

D_5 is not used.

D_4, B, indicates that a BREAK (BRK) instruction has been executed.

D_3, D, is the Decimal arithmetic mode flag. When **0**, arithmetic addition or subtraction is carried out in binary arithmetic, when **1**, in Binary Coded Decimal (BCD) arithmetic.

D_2, I, is the Interrupt disable flag. This will be discussed in Chapter 7.

D_1, Z, is the Zero result flag; it indicates whether the data result of instruction execution had all bits false $(D_1 = 1)$ or not $(D_1 = 0)$.

D_0, C, is the Carry bit. In arithmetic operations, it is used as the carry input and output (Chapter 3). It is also used in other logical operations.

6.2   The addressing modes of the 6502

Many microprocessor instructions require data transfers during their execution. The architecture of the microprocessor is designed to allow several very different methods by which the source and destination of the data are defined. This aspect of instruction execution is called the *addressing mode*; it was discussed briefly in Chapter 5.

The complexity of the IDL is reduced if all of the instruction interpretation sequences share common blocks of the control logic; this is achieved by having all of the instructions share common addressing modes. A study of the instruction set can therefore be divided into a study of the

Figure 6.2   The 6502 P Register.

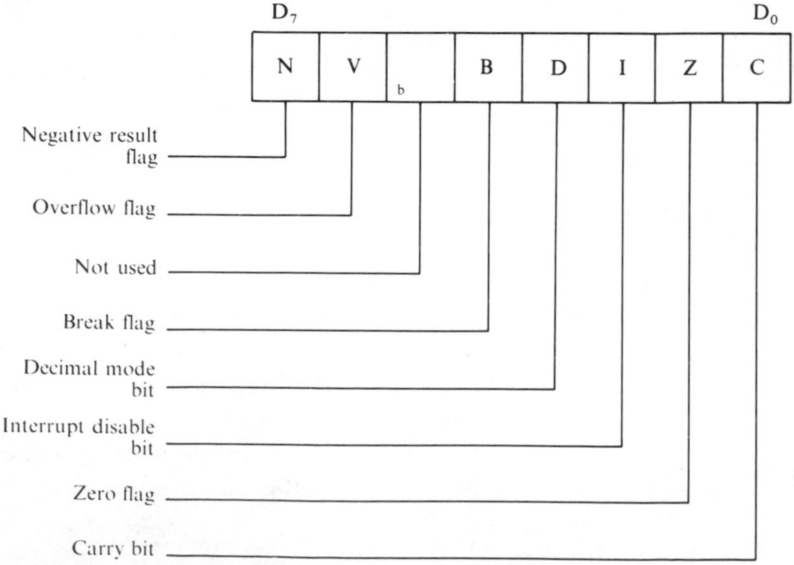

logical operations carried out by the various instructions and a study of their common addressing modes. In this section we will systematically explore the full range of addressing modes provided in the 6502. The control sequence required for each of these will not be detailed; instead, they will be described in terms of the simple program model of microprocessor operation. Each addressing mode will be described individually, to show how each instruction's operand is used to define the address or the argument to be used. Typical applications of each mode will be described where appropriate. A fuller description of their significance will be left until program examples are described later in this chapter.

In the diagrams below,

OPCODE stands for an instruction opcode,

d or a stands for a hexadecimal digit,

$dd stands for an 8-bit data word,

$aa stands for an 8-bit address (PAGE #0),

$aaaa stands for a 16-bit address.

6.2.1   Direct addressing modes

In all of these modes, the address is defined directly by the opcode or by the operand.

6.2.1.1   *Implied and Accumulator addressing modes*

In these two modes the opcode itself defines the source and/or destination to be used by the instruction. No operand is required; the full instruction uses only one word in memory.

MNEMONIC: OPCODE (single-word instruction)

EXAMPLE: TAX (Transfer contents of A Register into X Register)

SYMBOL: (none)

Figure 6.3   The IMPLIED addressing mode.

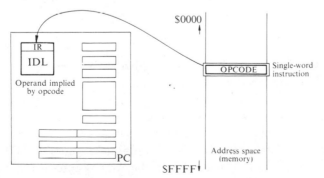

Some of the instructions for multi-bit logical manipulation of a word can be performed on the contents of the Accumulator. This is called Accumulator mode.

MNEMONIC: OPCODE (one-word instruction)

EXAMPLE: ROL A (Rotate the A Register contents)

SYMBOL: A

The operand designation (A) is not always specified in the mnemonic representation.

6.2.1.2 *Immediate addressing mode*

After the IDL has fetched the instruction opcode into the IR, the IDL has incremented the PC; it points (immediately) to the location of the data. The operand contains one word which is the data or argument.

MNEMONIC: OPCODE #$dd (two-word instruction)

EXAMPLE: LDA #$A0

SYMBOL: IMM

(The # sign specifies that the operand is the argument.)

Figure 6.4   The relationship between the opcode and the operand for IMMEDIATE addressing mode.

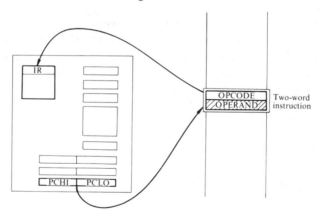

This mode is seldom used for anything other than accessing a constant embedded in the instruction.

6.2.1.3 *Zero page direct addressing mode*

The 6502 offers a fast, compact addressing mode when the accessed address is on PAGE #0. In this mode, after the IDL has fetched the opcode, the operand is fetched into the ADLO byte of the Address

Register and the ADHI byte is set to $00. The contents of the Address Register are then used to specify the address to be used by the instruction. Obviously, only a Zero Page (ZP) address can be accessed. (When the instruction execution is finished, the PC designates the address of the next instruction opcode.)

> MNEMONIC: OPCODE $aa (two-word instruction)
> EXAMPLE: LDA $8B
> SYMBOL: ZP

Figure 6.5 ZERO PAGE DIRECT addressing mode.

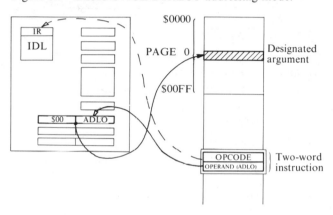

This mode is very fast and has a very short instruction; since it can only use 256 locations, it must be used sparingly and only when maximum speed and/or very compact code is needed.

6.2.1.4 *Absolute addressing mode (DIRECT)*

After the IDL has fetched the opcode into the IR, the PC is pointing at the first of two words which together represent the address to be used. The IDL fetches the first into ADLO and the second into ADHI; the address thus assembled in the Address Register is then used in the instruction execution.

> MNEMONIC: OPCODE $aaaa (three-word instruction)
> EXAMPLE: STA $7008
> SYMBOL: ABS

Figure 6.6   ABSOLUTE DIRECT addressing mode.

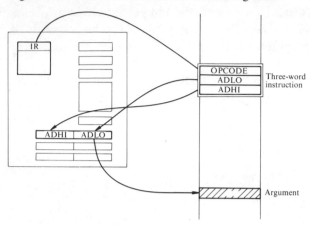

This mode is used for accessing data at an address remote from the program. It allows simple communication of variable data between program segments via a location (a 'pigeon-hole' or 'mailbox') which is defined and specified in *absolute* terms.

6.2.2   Indexed direct addressing modes

The X and Y Index Registers are intended principally to provide enhanced addressing modes; the IDL is able to use the ALU to add the contents of the specified index register to the Address Register contents, to specify the address to be used. The two direct modes described below thus allow a table to be set up in memory with its initial element *pointed* by the operand address. Any element of this table can then be selected by the offset value held in the index register.

6.2.2.1   *Zero page indexed addressing mode* (*DIRECT*)

This mode is a modification of the ZERO PAGE mode. The Address Register is loaded as in that mode (Section 6.2.1.3) but, before it is used to access an address, the contents of the defined index register is added to it. The IDL performs this with the ADHI held at $00; all elements of the indexed table thus must be on PAGE #0. If the addition overflows, the result will *rollover* within PAGE #0.

        MNEMONIC: OPCODE $aa, X
                           OPCODE $aa, Y (two-word instructions)
        EXAMPLES: LDA $3E, X
                       LDA, $D2, Y
        SYMBOLS: ZP, X
                   ZP, Y

Figure 6.7   ZERO PAGE X or Y INDEXED addressing mode.
(*a*) Note that the ALU is used to add the register contents to ADLO.
(*b*) The argument location diagram which depicts the procedure by
which the operand identifies the argument.

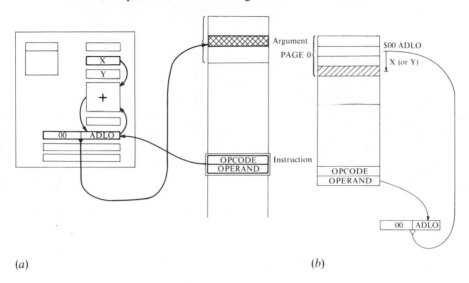

(*a*)   (*b*)

6.2.2.2   *Absolute indexed addressing modes (DIRECT)*

This mode, like the ZERO PAGE indexed mode, is a modification
of the ABSOLUTE mode described in Section 6.2.1.4. The Address
Register is loaded in similar manner with the two operands but, before the
assembled address is used, the content of the index register is added to
designate the *forwards indexed* address. A full 16-bit addition is carried
out; if the addition of ADLO and the index register overflows, the IDL
inserts another cycle to increment ADHI by one. In other words, page
crossing is allowed, but the IDL requires one more cycle for execution.

This mode allows a table of up to 256_{10} arguments to be established
anywhere in the address space and to be accessed by a single (indexed)
instruction.

MNEMONIC: OPCODE $aaaa, X
                     OPCODE $aaaa, Y (three-word instructions)
EXAMPLES: LDA $0308, X
                     STA $C0D8, Y
SYMBOLS: ABS, X
                     ABS, Y

Figure 6.8   ABSOLUTE X or Y DIRECT INDEXED addressing
modes: the extra cycle required for page crossing is shown dashed.
(*a*) The argument location diagram (*b*) depicts the procedure for
identification of the argument.

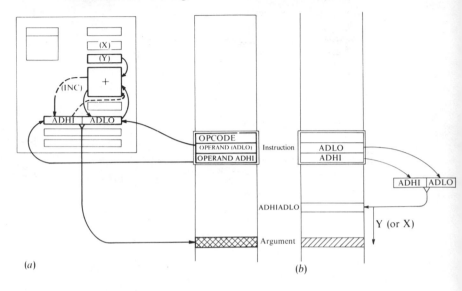

(*a*)   (*b*)

6.2.3   Indirect addressing modes

The instruction interpretation logic sequences described in
Chapter 5 can be extended to provide quite elaborate addressing modes.
An important example is for *indirect addressing modes*. As the name
implies, the operand is not used immediately (directly) to specify the actual
address of the argument; instead, it is used *indirectly* to specify the address
which 'points' to the argument! This sounds complicated, but the control
sequences are quite simple. In the first steps of the execution, the operand is
used as described in the DIRECT modes above, to assemble a 16-bit
address in the Address Register. But then, instead of using that address for
the argument's location, it is used to collect and assemble a further 16-bit
address which points to the argument. In other words, the first assembled
address *points* to the argument's address.

Some processors allow two-level, indirect addressing where the indirect
defined address is itself a pointer to the actual location to be used. (More of
this in Box 6.1.)

The 6502 has a very limited range of indirect addressing modes. It has a
single *absolute* indirect addressing mode used by only one instruction, and
has only two indirect *indexed* addressing modes, both of which use single-
word operands referring to addresses on PAGE #0. Being indirect modes,

they do not use the ZERO PAGE location as the actual argument address; instead, they use the information stored there to assemble an INDEXED address elsewhere in memory. These two modes use the ZERO PAGE information in quite different ways.

The control sequences generated by the IDL to carry out these two modes are rather involved and are left as exercises. In the description below, only the overall effect is described and depicted.

6.2.3.1 *Absolute addressing mode (INDIRECT)*

This addressing mode is an elaboration of the absolute DIRECT mode. In the 6502 microprocessor, it is only available for the unconditional JUMP instruction (see later). When the address specified by the two operand words has been assembled in the Address Register, it is transferred into the PC (Fig. 6.9, Step (1)) and two more fetch cycles are carried out to assemble an INDIRECT address in the Address Register. This address is then again transferred to the PC and the next instruction opcode is fetched from there (Fig. 6.9, Step (2)). The net effect is to transfer program execution to the address referenced *indirectly* by the contents of the address (and the next address) specified by the two operand words. (This is sometimes called *deferred* addressing.) Note that the original content of the PC, the pointer to the prior program, is lost during the execution of this instruction.

      MNEOMINC: OPCODE ($aaaa) (three-word instruction)
      EXAMPLE: JMP ($8045)
      SYMBOL: (ABS)

Figure 6.9   ABSOLUTE INDIRECT addressing mode. The two stages (Steps (1) and (2)) are shown in (*a*) and (*b*). The overall procedure is summarised in the argument location diagram (*c*).

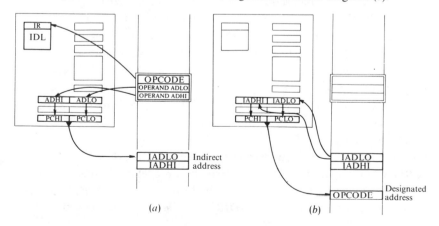

Figure 6.9 (c)

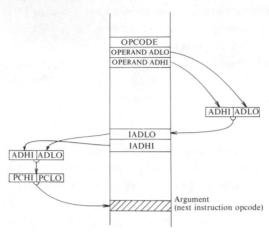

Note that the indirect action is usually symbolised by brackets around the operand, or by the @ sign, @ $aaaa.

This mode allows the destination of the JUMP to be redefined without modifying the operand of the JUMP instruction. This is very useful where many program segments contain JUMP instructions to the same destination. Without this addressing mode, a relocation of the target program would require that the operands in every JUMP instruction be changed; with this mode, it is only necessary to change the contents of the two addresses that specify the indirect address. (See Section 6.4.7.)

6.2.3.2   *Preindexed indirect zero page addressing mode*

The 6502 provides this mode only with the X Register as the index. It begins like the DIRECT ZERO PAGE INDEXED mode to *point* to an X-indexed location in PAGE #0. However, it then uses the information held in that address and the following address (rollover to location $0000 if necessary) to assemble a two-word (16-bit) address. This is the final designated address.

> MNEMONIC: OPCODE ($aa, X) (two-word instruction)
> EXAMPLE: LDA ($FA, X)
> SYMBOL: (ZP, X)

Note that this is a two-word instruction; it requires only the opcode and the ZERO PAGE operand. Note also that the bracketing of the mnemonic symbolises that the indexing is performed on the ZERO PAGE address before the result is used as an INDIRECT reference (preindexed).

Figure 6.10   PREINDEXED INDIRECT ZERO PAGE (X)
addressing mode. Only the argument location diagram is shown.

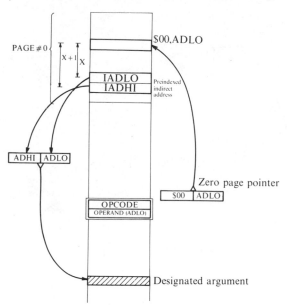

This mode allows simple reference to an address table set up in
PAGE #0. The value of the X Register uses this table to select argument
locations defined anywhere in memory.

6.2.3.3   *Post-indexed indirect zero page addressing mode*

This addressing mode can only use the Y Index Register. It begins
by using the operand to define a ZERO PAGE address. The contents of this
location and the following locations are used (rollover to $0000 if
necessary) to specify an INDIRECT address anywhere in memory. When
this INDIRECT address has been assembled, the contents of the
Y Register are added to it (page crossing is allowed) to designate the final
address.

　　　MNEMONIC: OPCODE ($aa), Y (two-word instruction)
　　　EXAMPLE: LDA ($8A), Y
　　　SYMBOL: (ZP), Y

Note that this is a two-word instruction. It requires only the opcode and a
single operand, the first ZERO PAGE address used. The use of the
Y Register as an index *after* the table starting address has been assembled is
clearly indicated by the position of the brackets.

Figure 6.11   POSTINDEXED INDIRECT ZERO PAGE (Y) addressing mode. Only the argument location diagram is shown.

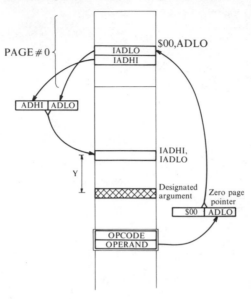

This mode allows a contiguous table of up to 256_{10} arguments to be specified anywhere in memory with the starting location defined by two words in PAGE #0. Any element of the table can be selected by the Y Register and can be accessed by a simple two-word instruction. Different tables can be selected by changing the contents of the PAGE #0 locations used.

6.2.4   The stack pointer – register addressing

Many microprocessors and all processors in large computers provide a class of addressing modes where the general-purpose internal registers are used as pointers to the data locations in memory. They can be in DIRECT REGISTER mode where the register contents are the actual address, or in INDIRECT (or DEFERRED) REGISTER mode where the register contents points to the locations which hold the address of the locations to be used. This class of addressing mode requires that either the internal registers have the same number of bits as the address bus, or that some means are provided to augment the register information to the full address width.

The 6502 only provides this addressing mode with the specialised S Register or Stack Pointer which creates the STACK. This is a temporary store for *automatic* placement and recovery of data words in the memory space of the computer; it provides a quick, compact repository for temporary data without the need for the program to keep track of where it is stored. It behaves like a pile or stack of paper. Information is added by placing it on top, and is recovered in reverse order by taking it from the top (Fig. 6.12(*a*)). Expressed in logic gates, it is called a *LIFO* (*Last In, First Out*) *REGISTER*. It is easily implemented within the instruction interpretation logic of the processor. It only requires the addition of an automatic increment or decrement of the Stack Pointer Register each time that it is used. The 6502 performs this as described below.

When a word of data is placed (*pushed*) onto the STACK, the IDL forces the S Register contents onto the address bus and then writes the data word to the pointed address; after this has been done, it decrements the contents of the S Register by one and thus points to the next lower address in memory (Fig. 6.12(*b*)). When a word of data is recovered (*pulled*) from the stack, the IDL first increments the S Register contents by one and then reads data from the address now specified by it (Fig. 6.12(*c*)). The special addressing modes are called REGISTER POSTDECREMENT and

Figure 6.12   (*a*) The STACK concept, (*b*) pushing a word onto the 6502 STACK (Step (1) – data is pushed to a location specified by S; Step (2) – decrement S ready for next push) and (*c*) pulling a word back (Step (1)   increment S to point to last data; Step (2) – fetch data).

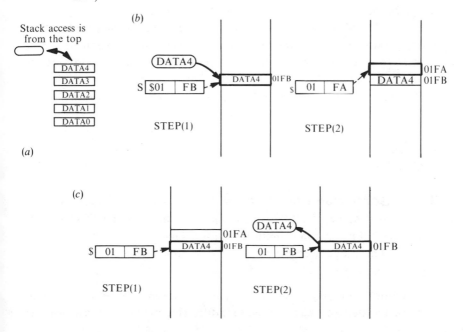

REGISTER PREINCREMENT respectively. They are usually symbolised thus:

Postdecrement on Register r: $(\mathbf{r})^-$

Preincrement on Register r: $^+(\mathbf{r})$

where the position of the sign shows when the change is made and, as before, the brackets indicate indirect use of the register contents.

Note that, in the 6502, only the low byte of the Stack Pointer can be manipulated, and that it can use only one page of memory. When accessed, the top 8 bits of the address bus are forced to $01. When the S Register overflows or underflows it does a rollover to the far end of PAGE # 1 *and gives no warning*. (More about this later!)

Many processors provide powerful stack facilities (see Box 6.1). Extended stack facilities can be generated by software (see Box 6.6).

6.2.5   Relative addressing modes

The concept of relative addressing modes is similar to that of indexed addressing. The location in memory that is to be used is specified by its displacement from another known location. Some processors provide relative addressing to enhance the addressing modes for instruction argument designation, but all use the relative addressing mode for manipulation of the PC contents. This simple enhancement of the instruction interpretation sequence provides an efficient means for modification of a program's execution sequence.

Unlike the indexed addressing modes, where the final designated address is determined by adding the index as an *unsigned* integer onto the table pointer to give a forwards *offset*, the relative addressing mode uses the instruction's operand as a *signed* integer value and thus provides for backwards or forwards offset. When applied to the PC, this forces the next instruction to be fetched from a remote location; program execution is transferred to a remote routine. In the 6502, this mode is only provided with the conditional branch instructions. (Section 6.4.6 will describe this mode in detail.)

6.3   Some general remarks about addressing modes

The range of addressing modes provided in the 6502 is very restricted, but it is a good compromise for a simple machine with an 8-bit opcode. The range seems a bit bizarre at first but, with practice and experience, efficient address handling can be found for most situations.

The major shortcoming of the 6502 is the absence of 16-bit, general-purpose registers that can be used for addressing memory. It is a *memory-*

oriented microprocessor; it favours logic operations to be carried out between an internal register (usually the Accumulator) and data in the memory space of the machine. This only requires registers of the databus width. The 8080, Z80 and their derivatives are all *register-oriented microprocessors*; they favour logic operations upon the contents of the internal registers. These can be treated as 16-bit words, and a wide range of 16-bit logic operations can be carried out after the registers have been loaded with the necessary information. The 16-bit registers, being the same width as the address bus, can be used as general-purpose address pointers; the instruction set is therefore enhanced by a range of powerful register addressing modes (see Box 6.1).

The 256_{10} possible codes of an 8-bit opcode are unable to provide a wide range of instructions with varied addressing modes. The Zilog Z80 and the Motorola 6809 have a considerable range of addressing modes with good register modes, but both require many two-word opcodes to represent them. The 6809, for example, requires separate opcodes and postbyte opcodes for all indexed addressing modes. The opcode specifies the logic operation and the class of addressing mode; if an indexed mode is selected, the postbyte opcode specifies the register to be used and the register mode to be selected. The resulting complexity of 6809 machine code makes it mandatory to use an assembler to write all but the simplest programs.

Box 6.1
Register addressing modes

If a processor has general-purpose internal registers of address bus width, they can be used as either data registers or as address pointers. The instruction set will invariably provide a range of *register* addressing modes. This box describes the common types of modes and shows some of their versatility and power.

Simple register addressing modes
IMMEDIATE: this mode has been described with the 6502. The register is used as the source or destination of data (or the argument) (Fig. 6.13(*a*)).

DIRECT: in this mode, the *contents* of the register are used to specify the *address* to be used as the source of destination of data (Fig. 6.13(*b*)).

INDIRECT: here the contents of the register are used to specify the address which holds the address to be used as the source or destination of data (Fig. 6.13(c)).

INDIRECT DEFERRED: this mode adds one more level of indirect access to the indirect mode; the register specifies the address which holds the address pointing to the address to be used (Fig. 6.13(d)).

The inherent power of the register addressing modes is that extra information is held within the registers of the processor; this information can be accessed very compactly and quickly by designating the register rather than an external address. If the arguments of an operation are in memory, two full address words must be specified, each requiring data transactions. If, however, the arguments are held within the processor's registers, they can be accessed by a single instruction. If there are eight registers, any one of them can be specified by a 3-bit code within the instruction opcode itself. With a 16-bit machine, the 16-bit opcode has plenty of space to define two register codes; all transaction information can thus be specified in a single instruction word using one register's contents as the source operand and another as the destination operand. The advantage of this is negated if the registers must first be loaded with information before the operation is executed. If, however, the program is working through large data arrays (indirect mode) or large data-pointer arrays (indirect deferred mode), the array address need only be loaded once. Thereafter, the register contents can be manipulated to access the various elements of the data arrays. This powerful capability is enhanced by the provision of auto-increment and auto-decrement options to the register addressing modes. These options allow the registers to scan automatically through tables.

Fig. 6.14 depicts some of the typical addressing modes provided in 16-bit *register-oriented* processors. The instructions depicted all would have single-word opcodes specifying the two registers.

Figure 6.13   Simple register addressing modes, (a) IMMEDIATE, (b) DIRECT, (c) INDIRECT and (d) INDIRECT DEFERRED. The arguments are shaded.

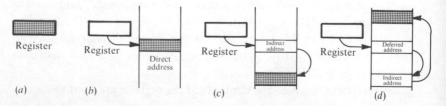

Figure 6.14   Typical register addressing modes for array processing:
(*a*) an immediate register source and a postincremented indirect
destination; repeated execution of this instruction would fill an array
with the constant in R0; (*b*) a predecremented source pointer and a
postincremented destination pointer; repeated execution allows an
array to be reversed; (*c*) an indirect postincremented source pointer
and a postincremented indirect deferred destination; this would
distribute the elements in the array pointed by the source register, to
the addresses specified in the array pointed by the destination register.

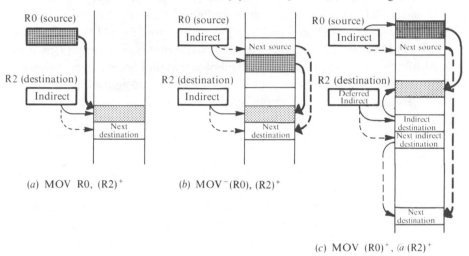

(*a*) MOV R0, (R2)$^+$ (*b*) MOV$^-$(R0), (R2)$^+$

(*c*) MOV (R0)$^+$, @(R2)$^+$

Box 6.1 shows some of the power and versatility, possible within
conventional computer architecture, to locate the arguments for logic
operations. The next section of this chapter returns to the simple 6502
microprocessor to describe the range of logic operations provided within
its instruction set. We will see that the simple architecture of the computer
allows a small range of instructions to achieve a very wide range of
operations. In a well-designed machine, the instructions are complete and
are able to emulate any logic task that a programmer can devise. The only
limits to the application of the machine are the time that the task requires
and the amount of program and logic variable space that it needs.

6.4   The instruction set of the 6502

The format of instructions for the 6502 is very simple. The first
word of an instruction is called the *opcode*; it defines the instruction
interpretation sequence and thus the logic operation and the addressing
mode used. It also defines the number of words in the remainder of the
instruction. This will be zero, one or two further words. These further

words are called the *operand* of the instruction and specify, with the addressing mode, the data or source to be used as the *argument* of the logic operation. Some logic operations in the instruction set require neither operand nor argument, some have no operand but have an argument, and some have an operand and two arguments. All are easily implemented by simple instruction execution logic.

A loose convention has been developed to symbolise the effect of instruction execution and to show the flow of data within the machine. The bold letters **A**, **X**, **Y** are used to represent the contents of the registers and the bold letter **M** is used to represent the contents of a memory element. Brackets around the bold letter symbolise that the contents are a pointer to an address etc. Arrows are used to show the movement of data between elements and logic operation symbols are used to symbolise multi-bit variable operations between elements. Here are three examples.

> **A** → **M** represents transferring the contents of the Accumulator to a memory element.
>
> **A** ← **(PC)** represents moving the data addressed by the PC into the Accumulator.
>
> **A** ← **A**$_x$ · **M**$_x$ represents taking the bitwise logical **AND** operation between the Accumulator and a memory location and putting the 8-bit result into the Accumulator.

We will use this convention throughout the rest of this chapter to depict the effect of instructions.

The information given in this section is brought together in concise, abbreviated form in Table 6.1 (p. 182). Refer to this table as you work through this section. It is advisable to read Section 6.6 and to gain access to machine language programming on a suitable computer (Apple, BBC etc.). You will then be able to try the short examples as you work through the text.

6.4.1   Data transfer instructions

We will start by describing the instructions which provide data movements between elements of the machine. There are two classes of data movement in the 6502. These are

(1) transfers between internal processor registers,

(2) transfers between an internal register and an external address.

The 6502 master/slave structure does not allow data to be moved directly between slave elements, such transfers must be done via an internal register.

6.4.1.1 *Internal data transfers*

All of these instructions use IMPLIED addressing and therefore need no operand.

(i) Four instructions provide data transfers from the accumulator to an index register and vice versa. Instruction execution leaves the data source unchanged. All four instructions affect the N and Z flags of the P Register as described in Section 6.1. The other status flags are not affected. Note that there are no instructions to swap data between registers or to move data directly between the index registers.

Mnemonic	Opcode	Operation	Description
TAX	$AA	$X \leftarrow A$	Transfers A contents to X
TXA	$8A	$A \leftarrow X$	Transfers X contents to A
TAY	$A8	$Y \leftarrow A$	Transfers A contents to Y
TYA	$98	$A \leftarrow Y$	Transfers Y contents to A

(ii) The contents of the Stack Pointer (S Register) can be moved (copied) into the X Register and vice versa. Moving the S Register contents into the X Register affects the N and Z flags of the P Register, but moving the X Register contents into the S Register has no effect on the flags.

Mnemonic	Opcode	Operation	Description
TSX	$BA	$X \leftarrow S$	Copies S into X (affects N, Z)
TXS	$9A	$S \leftarrow X$	Copies X into S (affects no flags)

(iii) Four instructions allow transactions between internal registers and the STACK area of memory space. These are not really internal data transfers; they use indirect, auto-increment or -decrement register addressing via the S Register. They are included here since they are single word instructions for data transfers.

Two instructions allow the Accumulator or the P Register to be *pushed* onto the stack; neither of these affect the P Register flags.

A further instruction allows a word to be *pulled* from the stack and put into the Accumulator (no flags are affected). The fourth instruction allows a word to be pulled off the stack and put into the P Register. This obviously affects all flags!

Mnemonic	Opcode	Operation	Description
PHA	$48	$(S)^{-} \leftarrow A$	Push A onto STACK
PHP	$08	$(S)^{-} \leftarrow P$	Push P onto STACK
PLA	$68	$A \leftarrow {}^{+}(S)$	Pull A from STACK
PLP	$28	$P \leftarrow {}^{+}(S)$	Pull P from STACK

Box 6.2(a)
Some examples

Example 6.1
There are no instructions for moving data between the two index registers. Such transfers must be done via the Accumulator or memory; the contents of the Accumulator may, however, be needed in a subsequent instruction and therefore must first be saved. The STACK is very convenient for this. The procedure for copying the X Register into the Y Register is depicted below in symbols, mnemonics and in *machine code*.

Operation sequence

$(S)^- \leftarrow A$

$A \leftarrow X$

$Y \leftarrow A$

$A \leftarrow {}^+(S)$

Code and notes

```
PHA    $48    :Push the A onto the STACK
TXA    $8A    :Transfer X contents into A
TAY    $98    :Transfer A contents into Y
PLA    $68    :Restore A off the STACK
```

Example 6.2
The 6502 provides no direct instructions for swapping register contents. The following routine allows the X and Y Register contents to be interchanged.

Operation sequence

$(S)^- \leftarrow A$

$A \leftarrow Y$

$(S)^- \leftarrow A$

$A \leftarrow X$

$Y \leftarrow A$

$A \leftarrow {}^+(S)$

$X \leftarrow A$

$A \leftarrow {}^+(S)$

Code and notes

```
PHA     $48     : Save A contents on the STACK
TYA     $8A     : Fetch Y into A
PHA     $48     : Push onto STACK
TXA     $8A     : Fetch X into A
TAY     $A8     : and then put into Y
PLA     $68     : Recover Y contents from STACK
TAX     $AA     : and put it into X
PLA     $68     : and finally, restore original A
```

It can be seen that, although internal swaps can indeed be carried out, some can be rather tedious. This demonstrates the difference between a logically *complete* set and a *full* set of instructions.

6.4.1.2 *Internal/external data transfers*

The six operations in this subset move data between an internal register and a memory element, leaving the data source as it was. The Accumulator is the favoured internal register (it has the best range of addressing modes), but the X and Y Registers can be used. All instructions which transfer data into an internal register affect the N and Z flags of the P Register, but instructions that move data out of the internal registers into memory have no effect. (See Table 6.1, p. 182.) The six operations with all possible addressing modes represent 31 different instructions.

The three operations to load internal registers are

LDA **A ← M**   LoaD the Accumulator (from memory)

LDX **X ← M**   LoaD the X Register (from memory)

LDY **Y ← M**   LoaD the Y Register (from memory)

Each operation has a different range of addressing modes. The LDA operation has eight different instruction opcodes defining addressing modes thus

LDA #$dd	–IMM–	LDA $aa	–ZP–
LDA $aaaa	–ABS–	LDA $aa, X	–ZP, X–
LDA $aaaa, X	—ABS, X—	LDA $aaaa, Y	–ABS, Y–
LDA ($aa, X)	–(ZP, X)–	LDA ($aa), Y	–(ZP), Y–

(Refer to Table 6.1 for opcodes etc.) The LDX operation has five different instruction opcodes

LDX #$dd	–IMM–	LDX $aa	–ZP–
LDX $aaaa	–ABS–	LDX $aa, Y	–ZP, Y—
LDX $aaaa, Y	–ABS, Y–		

The LDY operation also has five different instructions

LDY #$dd	–IMM–	LDY $aa	–ZP–
LDY $aaaa	–ABS–	LDY $aa, X	–ZP, X–
LDY $aaaa, X	–ABS, X		

The three operations to transfer (copy) data from the internal registers to memory space are

STA $A \rightarrow M$ STore the Accumulator in memory
STX $X \rightarrow M$ STore the X Register in memory
STY $Y \rightarrow M$ STore the Y Register in memory

The Accumulator instruction is favoured with seven addressing modes (seven different opcodes), while the X and Y Register instructions have only three addressing modes each.

STA $aa   –ZP–
STA $aa, X   –ZP, X–
STA ($aa), Y   –(ZP), Y–
STA ($aa, X)   –(ZP, X)–
STA $aaaa   –ABS–
STA $aaaa, X   –ABS, X–
STA $aaaa, Y   –ABS, Y–

STX $aa –ZP–   STX $aaaa –ABS–   STX $aa, Y –ZP, Y–

STY $aa –ZP–   STY $aaaa –ABS–   STY $aa, X –ZP, X–

The 6502 provides no instructions to transfer data directly between memory elements; this must always be done via an internal register.

Box 6.2(b)
More examples

Example 6.3
This example shows a short routine to move data at location $4000 to location $4180.

Transfer sequence

$A \leftarrow M1$

$M2 \leftarrow A$

Code and notes

```
LDA $4000   AD 00 40   : Copy $4000 contents to A
STA $4180   8D 80 41   : Copy A to $4180
```

Example 6.4
To load the second element of an array starting at $5000 with data $FE. (Refer to Table 6.1 (p. 182) for opcodes.)

Transfer sequences

X ← $01

A ← $FE

M ← **A**

Code and notes

```
LDX #$01     A2 01    : Load X to point to second element
LDA #$FE     A9 FE    : Load A with data
STA $5000,X  9D 00 50 : Copy A to memory
```

Example 6.5

A routine to swap the X and Y Register contents. (Refer to Table 6.1 for times etc.)

Transfer sequence

M1 ← **X**

M2 ← **Y**

  Y ← **M1**

  X ← **M2**

Code and notes

```
STX $FE     86 FE     : transfer X to Page#0
STY $FF     84 FF     : and Y also
LDY $FE     A4 FE     : transfer X data into Y
LDX $FF     A6 FF     : and Y data into X
```

6.4.2   Processor status instructions

Single-word instructions are provided to modify the contents of the Processor Status Register (P Register). Only the seven that are logically necessary are provided. All use implied addressing mode.

CLC	CLear C bit (to **0**)
SEC	SEt C bit (to **1**)
CLD	CLear D flag (binary arithmetic mode)
SED	SEt D flag (BCD arithmetic mode)
CLI	CLear Interrupt disable flag (enable interrupts)
SEI	SEt Interrupt disable flat (disable interrupts)
CLV	CLear arithmetic oVerflow flag

(The purpose of the Interrupt Flag control will be explained in Chapter 7.)

6.4.3 Multi-bit word logical operations

A set of seven simple multi-bit logical operations are provided in the instruction set; these are sufficient to allow any bit of any word in the memory space to be modified in all possible ways. They classify into two types, with single or double arguments.

The four single-argument operations allow the bits of a single word to be left- or right-shifted one place; the bit shifted out of the word is placed into the C bit of the P Register. The bit shifted into the word can either be a logical **0** (ASL, LSR), or can be the prior content of the C bit (ROL, ROR).

The logic operation is actually carried out within the ALU; the IDL transfers the argument to the ALU, commands it to carry out the operation and then sends the result back to its source.

ASL　　Arithmetic Shift Left (cf. multiply by 2)

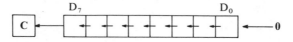

LSR　　Logic Shift Right (complement of ASL)

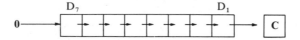

ROL　　ROtate Left via carry bit (a 9-bit rotation!)

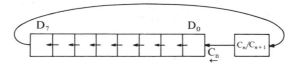

ROR　　ROtate Right via carry bit

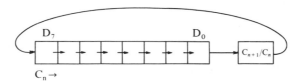

All four operations are provided in the same range of addressing modes
　　　　OPC A (Accumulator)
　　　　OPC $aaaa
　　　　OPC $aa
　　　　OPC $aa, X
　　　　OPC $aaaa, X

The instruction interpretation sequences make the A mode very fast (2 cycles) and all other modes slow (5–7 cycles). All operations affect the N and Z flags (LSR will always clear N), and the C bit is always loaded with the old D_7 (ASL, ROL) or the old D_0 (LSR, ROR).

These operations are very important in modelling arithmetic multiplication and division (recall Chapter 3), but are also of great use in moving bits of a word into more favourable positions for other operations.

The remaining three operations of this group use two arguments, one of which is always in the Accumulator. Each carries out a multi-bit, bitwise logical operation between the Accumulator and a specified memory location, and puts the bitwise result into the Accumulator.

All logic operations are performed within the ALU unit of the 6502. The IDL transfers the data specified by the operand into the ALU, presents the Accumulator contents to the ALU, commands the ALU to carry out the required operation and finally returns the result to the Accumulator (Fig. 6.15).

> **AND** $\quad A_x \leftarrow A_x \cdot M_x$
>
> Carries out a multi-bit logic **AND** between the Accumulator and memory, the result returned in the Accumulator.
>
> **ORA** $\quad A_x \leftarrow A_x + M_x$
>
> Carries out a multi-bit logic **OR** between the Accumulator and memory, the result returned in the Accumulator.

Figure 6.15 Instruction interpretation of logical operations via the ALU. (*a*) Step (1); fetch the designated memory data word into the ALU. (*b*) Step (2); present Accumulator contents to the ALU, execute the logic operation and latch the result. (*c*) Step (3); move result back to the Accumulator.

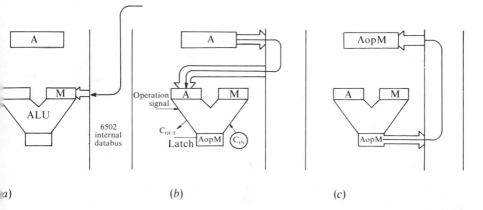

(*a*)   (*b*)   (*c*)

EOR $A_x \leftarrow A_x \oplus M_x$

Carries out a multi-bit **EOR** operation in similar manner.

All of these are provided in all possible addressing modes except PAGE #0, Y indexed. All affect only the N and Z bits of the P Register.

These three simple instructions, used with data movement instructions, allow any pattern of bits within any word in registers or memory to be cleared, set or negated.

Box 6.2(c)
Examples

Example 6.6

The **AND** operation allows any desired bits of the word in the Accumulator to be cleared. If you are not convinced, draw up a truth table for the two-input **AND** operation and consider one input as a control or 'mask' bit. The short segment below forces the m.s.b. and the l.s.b. of the data at $5040 to be cleared.

Code and notes

```
LDA #$7E    A9 7E    :Set up a mask word in A (01111110)
AND $5040   2D 40 50 :Mask out the data word
STA $5040   8D 40 50 :Restore the masked word to memory
```

The effect of the **ORA** operation is to allow any bit(s) to be selectively set in a word; the **EOR** operation allows any bit(s) to be negated. These operations are left as exercises for the reader.

Example 6.7

There is no instruction to set the V flag in the P Register (and there is never a need to do it!). This little segment does it, nonetheless. Note that, even though we have no direct instruction to access the P Register, it can be reached using a sequence of other instructions, via the STACK.

Code and notes

```
PHP         08       :Push P onto STACK
PLA         68       :and fetch it into A
ORA #$40    09 40    :Set D6 in A
PHA         48       :and then transfer
PLP         28       :the new word back to P
```

Example 6.8

This little segment allows an 8-bit rotation of the word in the Accumulator. Compare this with the 9-bit rotation provided within the instruction set (ROR, ROL).

Code and notes

```
ASL A    0A    :A one bit left
PHP      08    :Save the C bit that fell out
LSR A    4A    :A back one bit
PLP      28    :restore the previous C bit
ROL A    2A    :and finish the 8-bit rotation
```

This simple little routine again shows how operations not explicitly provided in the instruction set can be created from sequences of available operations. This is the difference between a *complete* logic set and a *full* logic set; the trade-off is between machine simplicity and program code length/speed. The code shown executes in 14 cycles and requires 5 words. It is left as an exercise to attempt a faster or shorter 6502 routine.

6.4.4 Arithmetic operations

Instructions for single-argument and double-argument arithmetic operations are provided in the instruction set. The single operand instructions allow selected elements to be incremented or decremented; double-argument instructions perform normal arithmetic addition and subtraction.

The 6502 provides single-word instructions to increment or decrement the two index registers using IMPLIED addressing mode. They work correctly only in binary mode; decimal mode gives weird results.

Only the N and Z flags are affected on the result of the operation. Over/underflows are *not* flagged in the C or V bits; the registers just rollover.

$$\text{INX} \qquad X \leftarrow X +_{ar} 1$$
$$\text{INY} \qquad Y \leftarrow Y +_{ar} 1$$
$$\text{DEX} \qquad X \leftarrow X - 1$$
$$\text{DEY} \qquad Y \leftarrow Y - 1$$

These increment and decrement operations are also executable in memory space using PAGE#0, ABSOLUTE, PAGE#0,X and ABSOLUTE,X addressing modes. (The Accumulator cannot be incremented or decremented directly.)

$$\text{INC} \qquad M \leftarrow M +_{ar} 1$$
$$\text{DEC} \qquad M \leftarrow M - 1$$

Conditions and status flags are as for INX and DEX.

The 6502 provides two further operations which model addition and subtraction with integer numbers precisely in the manner described in some detail in Sections 3.2 and 3.3. The full-adder logic is included within the ALU as depicted in Fig. 3.18. Instruction interpretation is depicted in Fig. 6.16 (refer also to Fig. 6.15).

$$\text{ADC} \qquad A \leftarrow A +_{ar} M +_{ar} C_{IN}$$

The Accumulator contents are ADded to the designated memory contents with the C bit as carry-in, and the result is put into the Accumulator.

$$\text{SBC} \qquad A \leftarrow A +_{ar} \bar{M} +_{ar} C_{IN}$$

The Accumulator contents are added to the negated M contents with the C bit as carry-in, the result going into the Accumulator. If the C bit were set beforehand, this models the subtraction of M from A, hence the mnemonic 'SuBtract memory from accumulator with Carry'.

Both operations are provided in all possible addressing modes with the exception of PAGE#0, Y indexed. Both affect the N and Z flags in the normal way. The C flag is affected as described above. If the *signed number line* boundary is crossed by the result, the V flag will be set, otherwise it will be cleared. It thus never needs to be set or cleared by the program for proper execution of the operations; it is not a *bit* used in the operation execution, it is a status *flag*.

Figure 6.16   Add/subtract instruction interpretation in the 6502. (*a*) Step (1); with specified memory latched into the ALU, the add or subtract command is issued, the result is latched. (*b*) Step (2); the result is moved to the Accumulator and the Carry is latched to the C bit.

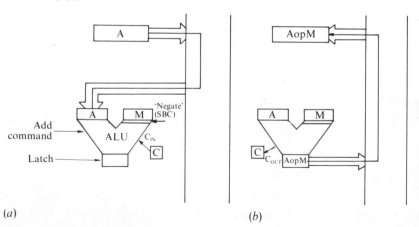

(*a*)   (*b*)

The C bit must be cleared (CLC) before addition (ADC) or set (SEC) before subtraction (SBC) of single word numbers.

The design of the addition hardware provides subtle and efficient multi-word arithmetic as well as a clever merging of addition and subtraction. Consider unsigned integer addition. If the addition of two words overflows the 8-bit number line, the C bit will be set, ready for a subsequent addition; if multi-word 'numbers' are added in ascending word-magnitude order, carry propagation is handled automatically.

Unsigned integer subtraction behaves in a similar manner. Recall that if two numbers are subtracted by using the two's-complement algorithm

$$\mathbf{A} - \mathbf{B} = \mathbf{A} +_{ar} \mathbf{\bar{B}} +_{ar} \mathbf{1}$$

then, if the result does not underflow (i.e. go negative), the addition will result in an overflow C_{OUT}. If an underflow does result, the addition result will be in two's-complement form (as a negative number) and there will be no C_{OUT} (Exercise 3.7). This allows multi-word numbers to be subtracted simply by carrying out single-word subtractions in ascending word-magnitude order with automatic *borrow* propagation.

Signed integers (two's complement) are only slightly more difficult to handle. The addition/subtraction operation for single-word numbers is as above, but an over/underflow condition is signalled by the V flag of the P Register. The addition and subtraction of multi-word numbers is handled automatically by normal carry propagation if the operation is carried out in ascending word-magnitude order; after the last (most significant) addition or subtraction, the V bit signals over/underflows of a signed number result.

When the D bit of the P Register is set low, the ALU performs arithmetic operations as if the data inputs are binary numbers. While the D bit is high, however (decimal mode), the ALU treats the data inputs as BCD numbers. This mode is provided more for historical reasons than for completeness; it is left to the reader to explore the orderly handling of signed decimal single- and multi-word numbers.

Box 6.2(d)
Examples

From here on, examples will be presented in a format that introduces the principal features of Editor/Assembler languages. They will also be located in memory as machine code suitable for loading and running.

Example 6.9

A short routine to compute the sum and difference of two data words.

We will need two input data words and two output words, one for the sum and one for the difference. We will ignore arithmetic over/underflow at this stage, but will return to it later. (This example uses unsigned integer numbers; conversion for signed integers is left as an exercise.)

Variable map

$4000	M1	First input data
$4001	M2	Second input data
$4002	SR	Sum result (M1 + M2)
$4003	DR	Difference result (M1 − M2)

Code and notes

```
LDA M1    $4010: AD 00 40    :Fetch one argument
CLC              18          :Clear C bit
ADC M2           6D 01 40    :Add arguments
STA SR           8D 02 40    :Save Sum
LDA M1           AD 00 40    :Fetch again
SEC              38          :Set C bit
SBC M2           ED 01 40    :Subtract
STA DR           8D 03 40    :Save Difference
BRK              00          :Break execution of segment.
```

(The BRK instruction BReaKs execution of this program; it will be discussed in Chapter 7.)

Example 6.10

This example shows how two-word unsigned integer numbers are subtracted using automatic carry (borrow) propagation.

Variable map

	HIBYTES		LOBYTES	
$4008	M1HI	$4009	M1LO	First two-word integer
$400A	M2HI	$400B	M2LO	Second two-word integer
$400C	DRHI	$400D	DRLO	Difference (**M1 − M2**)

Code and notes

```
LDA M1LO   $4020: AD 09 40    :Fetch 1st low word
SEC               38          :Set C bit first time
SBC M2LO          ED 0B 40    :Subtract
STA DRLO          8D 0D 40    :Store low difference
LDA M1HI          AD 08 40    :Fetch 1st high word
SBC M2HI          ED 0A 40    :Use previous C result !
STA DRHI          AD 0C 40    :Store high difference
BRK               00
```

6.4.5 Memory compare and test operations

The 6502 instruction set includes two operations that allow the contents of memory to be tested with, or compared with, the contents of an internal register. These operations do not alter the memory or register contents, they only effect the P Register flags or bits.

The COMPARE operation carries out an arithmetic subtraction of the designated memory address from the implied register with the C bit forced to **1** (SEC is not needed). The result is not returned to the register; it is not affected by the operation. The N, Z and C bits of the P Register are all affected in the same manner as for SBC, but note that the V bit is *not* altered. Note well that the C bit must be used as the test bit for unsigned integer comparison, and that the N flag will indicate the sign of the result after *signed* integer comparison.

$$\text{CMP} \qquad A +_{ar} \bar{M} +_{ar} 1$$

subtracts M from A, affecting only N, Z flags and C bit. This operation is provided in all possible addressing modes except PAGE #0, Y.

$$\text{CPX} \qquad X +_{ar} \bar{M} +_{ar} 1$$

As for CMP, but uses the X Register contents and provides only IMMEDIATE, PAGE #0 and ABSOLUTE modes.

$$\text{CPY} \qquad Y +_{ar} \bar{M} +_{ar} 1$$

As for CPX, but uses the Y Register contents.

The BIT (bit-test) operation is a hybrid. It transfers the contents of D_7 and D_6 of the specified memory into the N and V flags respectively of the P Register and performs an 8-bit logic comparison between the Accumulator and designated memory contents; the Z flag is set if they are the same, otherwise it is cleared.

$$\text{BIT} \quad M_7 \to N \quad (P_7); \qquad M_6 \to V \quad (P_6)$$
$$1 \to Z \quad (P_0) \quad \text{if } M = A,$$
$$0 \to Z \quad \text{if } M \neq A$$

These operations are intended as a test when an array of words is to be compared with, or tested against, constant(s) held in the internal registers. (This is why they do not change register contents.) They are also of some use for testing when a loop is finished. (See the following section.)

The BIT operation is well suited to testing the status of external devices; this will be discussed in Chapter 7.

6.4.6   Conditional branches

The 6502 uses the relative addressing mode to manipulate the contents of the PC. (This was introduced and discussed in Section 6.2.5.) The only application in the 6502 is in *conditional branching*, where the state of a designated flag in the P Register is used to decide whether or not the PC is to be modified. The effect of the modification is to fetch the next instruction from a location other than that following the current instruction. Conditional branching thus allows program execution either to continue to that instruction or to be *branched* elsewhere, dependent on the *condition* of a flag or bit in the P Register.

The 6502 uses a single-word branch operand which is treated as a signed integer and is added to the PC contents. (Page boundary crossing is made in either direction, but an extra cycle is needed.) The branch can therefore be $7F$ (127_{10}) words forwards in memory, or 80 (128_{10}) words backwards. At the end of the instruction execution cycle, the PC would normally be pointing at the following opcode; the branch is therefore made relative to that address. The branch must, of course, be to the opcode of the target instruction.

The 6502 provides instructions for conditional branching on either state of the N, Z, V flags or the C bit. If the specified condition is present, the instruction operand is added to the PC.

BPL	Branch if $N=0$
BMI	Branch if $N=1$
BVC	Branch if $V=0$
BVS	Branch if $V=1$
BNE	Branch if $Z=0$
BEQ	Branch if $Z=1$
BCC	Branch if $C=0$
BCS	Branch if $C=1$

A particular advantage of relative addressing is that the machine language program needs no absolute reference to the target locations in the program; it employs *position independent code* and will therefore run unmodified, anywhere in memory. The implications of this for modular programming are obvious.

Two serious shortcomings of the 6502 are the omission of *unconditional* branches, and that branches are limited to a single offset word. Most other microprocessors (including the enhanced 6502C used in the Apple IIC) provide both conditional and unconditional branching and provide the option of *short* (one-word) or *long* (two-word) branches. The problem of writing position independent code for the 6502 can be handled by specialised assembler programs which allow *relocatable code* to be created.

Box 6.2(e)
Examples

Example 6.11

The short routine below demonstrates the use of conditional branching. It is a time-delay routine which invokes nested loops; its execution time is a second-order polynomial of the argument fetched from the address $FF on PAGE #0.

Code, notes and flow chart

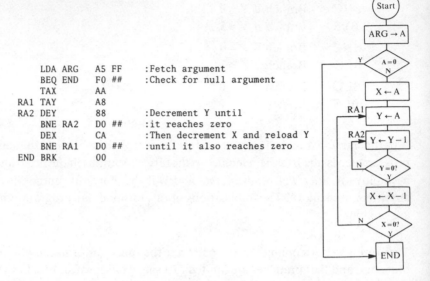

```
            LDA ARG    A5 FF    :Fetch argument
            BEQ END    F0 ##    :Check for null argument
            TAX        AA
     RA1    TAY        A8
     RA2    DEY        88       :Decrement Y until
            BNE RA2    D0 ##    :it reaches zero
            DEX        CA       :Then decrement X and reload Y
            BNE RA1    D0 ##    :until it also reaches zero
     END    BRK        00
```

The code on the far left is written in Assembler format with labelled locations. If coded directly by hand, the draft code (second table above) must be used to calculate the branch offsets. We will go through this carefully in this first example of branching.

When the processor reaches the first branch instruction, the PC will be pointing at the next OPCODE ($AA). The branch is to move the PC forwards eight words in memory to END ($00). Thus the offset must be $08.

The second branch instruction must move the PC back three words (from OPCODE $CA to $88), $FD in hexadecimal notation.

The third branch must move the PC back seven words, so that its operand must be $F9.

(Table 6.3 at the end of this chapter depicts all backward and forward relative branch operands.) The final position independent code is thus

```
            A5 FF
            F0 08
            AA
            A8
            88
            D0 FD
            CA
            D0 F9
            00
```

Derivation of the time-delay polynomial is left as an exercise.

Example 6.12

This example demonstrates several features. The routine runs through a table of 196_{10}($C4) data words whose starting address is pointed by PAGE#0 locations $FE, FF. The routine tallies the number of occurrences of each specific word pattern (ignoring $00 and $FF) in a page of memory ($4200–FF) specified within the program. This page is assumed to be clear at the start of the routine; the reader should extend the code to perform this.

Code, notes and flow chart

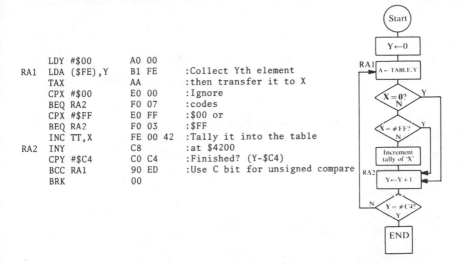

```
        LDY #$00        A0 00
RA1     LDA ($FE),Y     B1 FE       :Collect Yth element
        TAX             AA          :then transfer it to X
        CPX #$00        E0 00       :Ignore
        BEQ RA2         F0 07       :codes
        CPX #$FF        E0 FF       :$00 or
        BEQ RA2         F0 03       :$FF
        INC TT,X        FE 00 42    :Tally it into the table
RA2     INY             C8          :at $4200
        CPY #$C4        C0 C4       :Finished? (Y-$C4)
        BCC RA1         90 ED       :Use C bit for unsigned compare
        BRK             00
```

Note that the Y Register indexes through the scanned table, the X Register indexes into the tally table. Note also that the absence in the 6502 of indirect addressing for loading the X Register, and of indexing via the A Register, requires a data shuffle between A and X.

An inefficient method of ending the loop is used in this example. It is preferable, whenever possible, to end a loop when the index reaches zero. This yields a tidier and faster test, using the P Register status flags set by the increment or decrement operation. It also allows memory locations to be used efficiently as multi-word index counters.

6.4.7   Jumps and subroutines

The 6502 provides two types of instruction for unconditional transfer of program execution. One of these is a simple *jump*; the PC contents are overwritten by a new opcode address. It is provided in both absolute and indirect addressing modes. The other instruction pushes the old PC contents onto the STACK before the new address is entered. This is the JUMP to SUBROUTINE instruction; it is provided only in absolute addressing mode. It allows program execution to be returned, when the old PC contents are recovered from the stack, to the opcode that followed the subroutine jump instruction. This recovery is performed by another instruction called the RETURN from SUBROUTINE.

JuMP ABSOLUTE

JMP $aaaa   **PCLO ← M,   PCHI ← M+1**

Transfers program execution to the absolute address specified by the two-word operand.

JuMP INDIRECT

JMP ($aaaa)   **PCLO ← (M+1, M),   PCHI ← ((M+1, M)+1)**

Transfer program execution to the indirect address specified (see Section 6.2.3.1) by the two-word operand.

The 6502 has a 'bug' in this instruction. If the indirect address in the operand is at the end of a page (i.e. if $M = \$FF$) then, during assembly of the final destination address, the IDL 'forgets' to increment the page. The second address word is fetched from the start of the same page, not from the next page!

Jump to SubRoutine

JSR $aaaa   **(SP)⁻ ← PCHI**
              (SP)⁻ ← PCLO
              PC ← aaaa

The PC contents are pushed onto the stack and the two-word operand is moved into the PC. The next opcode is fetched from there. The use of the stack allows recursive and reentrant subroutine calls (see Box 6.3).

ReTurn from Subroutine

RTS   **PCLO ← ⁺(SP)**
        PCHI ← ⁺(SP)

The PC is loaded from the stack. *Provided that the Stack Pointer has not been misplaced*, program execution will resume at the instruction following the last encountered JSR.

ReTurn from Interrupt

RTI

This instruction will be described in the next chapter.

Box 6.3
The STACK, recursive subroutines and RPN

Early computers used one or more reserved registers to store the PC before the processor executed a subroutine jump. This limited the number of subroutines that could be nested. The use of a stack to store the return addresses avoids this problem. The LIFO organisation of the stack allows the processor to wend its way through multiple subroutine jumps (and even reentrant jumps into itself), limited only by the space available in the stack. As subroutine returns are encountered, the processor wends its way back in precise reverse order, provided, of course, that the user has not violated the stack pointer within the subroutines. Many processors now provide two or more stacks to avoid this; one is used by the system (i.e. for return addresses), the other for the user's software. Only *data* can be violated by the user; the processor will always be able to find its way back through the subroutines.

Reverse Polish Notation (RPN) for arithmetic equations uses the LIFO Stack concept. It provides a simple formalism for representing an equation as a string of arguments and operations, without brackets or punctuation. It is well suited to machine implementation; Hewlett-Packard use it for their hand-held calculators.

An operation is performed by pushing all its arguments onto a stack and then specifying the operation to be carried out on them. This scheme allows any arithmetic expression to be expressed as a stream of arguments and operations without punctuation or brackets. Consider, for example, the equation written conventionally as

$$(D+C)^{1/2}/(B+A)$$

In RPN this would be written as

A:B:add:C:D:add:sq. root:divide,

with arguments and arithmetic commands interspersed. The contents of the stack during execution of the stream are shown below. The stack pushes down ('↓' signifies a push) and the execution sequence is from left to right.

Opn.:	↓	↓	Add	↓	↓	Add	Sq. root	Divide
	A	B	B+A	C	D	D+C	$(D+C)^{1/2}$	$(D+C)^{1/2}/(B+A)$
		A		B+A	C	B+A	B+A	
					B+A			

Operations requiring one or two arguments from the stack are shown in the example but the scheme can be elaborated to any degree of complexity and to logical operations.

The twin concepts of recursive subroutines and parameter stacks are implemented in a *stack-oriented* programming language such as FORTH. The language defines a few simple operations which can be expressed in a small core of machine language code, but the structure of these operations allows them to be called by name (using a look-up table of their entry points) and to be nested into more complex procedures which can, in turn, be called by name. The scheme elaborates naturally into a multiple-branched, nested subroutine network. The core of the language is a software stack which allows the program to wend its way through the complex nested structure. The program is little more than a string of procedure names. Parameters needed in the subroutines are merged with the procedures stream in the same way as in RPN.

This concept of stack-based parameter/argument lists can be used to very good advantage in conventional languages such as BASIC, to simplify data transfers between a main routine and subroutines. The main program can push a string of parameters onto the main stack or onto a software stack before calling a subroutine; the subroutine need simply fetch its parameters from there without needing to know the location of the main routine and its parameters. Most 16-bit processors provide two stack pointers or stack-oriented register addressing, to make this easy.

6.4.8   Miscellaneous instructions

There are two instructions in the 6502 repertoire that have not fitted into the classification scheme above.

Null OPeration

$$\text{NOP} \qquad \text{PC} \leftarrow \text{PC} +_{ar} 1$$

This instruction simply fetches the next opcode. It is used principally to pad out programs either to allow space for future modification or to fill in shortened sections.

BReaK execution

$$\text{BRK}$$

This instruction forces program execution to a hardware defined address; it is called a *software interrupt* and will be discussed fully in Chapter 7.

6.5   A summary of the 6502 instruction set

Box 6.4 provides a convenient summary of the full instruction set
  of the 6502.

Table 6.1 depicts every instruction classified by type and referred
  to the section in this chapter where it is described.

Table 6.2 depicts all the opcodes in numerical order.

Table 6.3 is a backwards and forwards relative branch operand
  table.

Table 6.1. 6502 instructions by operational class

| Group | Mnemonic | Addressing modes and opcodes (byte count)/(cycle count)[a] | | | | | | | | | | | Status bits effected[c] | | | | | |
		Implied 1/2	Accumulator 1/2	Immediate 2/2	ABS 3/4	ABS,X 3/4[b]	ABS,Y 3/4[b]	Zero page 2/3	ZP,X 2/4	ZP,Y 2/4	IND,X 2/6	IND,Y 2/5[b]	N	Z	C	I	D	V
Transfers Section 6.4.1.1	TAX	AA											X	X	—	—	—	—
	TAY	A8											X	X	—	—	—	—
	TXA	8A											X	X	—	—	—	—
	TYA	98											X	X	—	—	—	—
	TSX	BA											X	X	—	—	—	—
	TXS	9A											—	—	—	—	—	—
	PHA	48(3)											—	—	—	—	—	—
	PLA	68(3)											X	X	—	—	—	—
	PHP	08(4)											—	—	—	—	—	—
	PLP	28(4)											X	X	X	X	X	X
Loads and saves (Section 6.4.1.2)	LDA			A9	AD	BD	B9	A5	B5		A1	B1	X	X	—	—	—	—
	LDX			A2	AE		BE	A6		B6			X	X	—	—	—	—
	LDY			A0	AC	BC		A4	B4				X	X	—	—	—	—
	STA				8D	9D(5)	99(5)	85	95		81	91(6)	—	—	—	—	—	—
	STX				8E			86		96			—	—	—	—	—	—
	STY				8C			84	94				—	—	—	—	—	—
Status modify (Section 6.4.2)	CLC	18											—	—	0	—	—	—
	CLD	D8											—	—	—	—	0	—
	CLI	58											—	—	—	0	—	—
	CLV	B8											—	—	—	—	—	0
	SEC	38											—	—	1	—	—	—
	SED	F8											—	—	—	—	1	—
	SEI	78											—	—	—	1	—	—

Mnemonic	Accumulator / Implied	Immediate / Relative	Absolute	Absolute, X	Absolute, Y	Zero page	Zero page, X	(Indirect, X)	(Indirect), Y	N	V	B	D	I	Z	C
Logic functions (Section 6.4.3)																
ASL	0A		0E(6)	1E(7)		06(5)	16(6)			X	—	—	—	—	X	D_7
ROL	2A		2E(6)	3E(7)		26(5)	36(6)			X	—	—	—	—	X	D_7
ROR	6A		6E(6)	7E(7)		66(5)	76(6)			X	—	—	—	—	X	D_0
LSR	4A		4E(6)	5E(7)		46(5)	56(6)			0	—	—	—	—	X	D_0
Binary arithmetic (Section 6.4.4)																
AND		29	2D	3D	39	25	35	21	31	X	—	—	—	—	X	—
EOR		49	4D	5D	59	45	55	41	51	X	—	—	—	—	X	—
ORA		09	0D	1D	19	05	15	01	11	X	—	—	—	—	X	—
ADC		69	6D	7D	79	65	75	61	71	X	X	—	—	—	X	X
SBC		E9	ED	FD	F9	E5	F5	E1	F1	X	X	—	—	—	X	X
INC			EE(6)	FE(7)		E6(5)	F6(6)			X	—	—	—	—	X	—
INX	E8									X	—	—	—	—	X	—
INY	C8									X	—	—	—	—	X	—
DEC			CE(6)	DE(7)		C6(5)	D6(6)			X	—	—	—	—	X	—
DEX	CA									X	—	—	—	—	X	—
DEY	88									X	—	—	—	—	X	—
Compares (Section 6.4.5)																
CMP		C9	CD	DD	D9	C5	D5	C1	D1	X	—	—	—	—	X	X
CPX		E0	EC			E4				X	—	—	—	—	X	X
CPY		C0	CC			C4				X	—	—	—	—	X	X
BIT			2C			24				M_7	M_6	—	—	—	X	—
Branches (Section 6.4.6)																
BCC		90								—	—	—	—	—	—	—
BCS		B0								—	—	—	—	—	—	—
BEQ		F0								—	—	—	—	—	—	—
BNE		D0								—	—	—	—	—	—	—
BMI		30								—	—	—	—	—	—	—
BPL		10								—	—	—	—	—	—	—
BVC		50								—	—	—	—	—	—	—
BVS		70								—	—	—	—	—	—	—

RELATIVE ADDRESSING: Requires 3 cycles if branch is on the same page, 4 if to adjacent page.

Continued overleaf

Table 6.1 (cont.)

Group	Mnemonic	Implied 1/2	Accumulator 1/2	Immediate 2/2	ABS 3/4	ABS,X 3/4[b]	ABS,Y 3/4[b]	Zero page 2/3	ZP,X 2/4	ZP,Y 2/4	IND,X 2/6	IND,Y 2/5[b]	N	Z	C	I	D	V
Jumps (Section 6.4.7)	JMP				4C(3)						(6C[d])		—	—	—	—	—	—
	JSR				20(6)								—	—	—	—	—	—
	RTS	60(6)	Stack is										—	—	—	—	—	—
	RTI	40(6)	used[e]										(From stack)[e]					
Others (Section 6.4.8)	NOP	EA											—	—	—	—	—	—
	BRK	00(7)	[e]										—	—	—	1	—	—

Addressing modes and opcodes (byte count)/(cycle count)[a]

Status bits effected[c]

[a] A bracketed number next to an opcode indicates the cycle count (if it is different from that at the head of the column).
[b] In these addressing modes, one more cycle is required if a page boundary is crossed.
[c] A dash (—) indicates that the operation has no effect on the Status Flags, X indicates that the flat is effected by the operation and 0 or 1 indicates a flag state that is forced by the operation.
[d] The operand is used as an indirect address pointer (Sections 6.2.5 and 6.4.7).
[e] Stack is used: refer to the relevant section.

Table 6.2. *6502 instruction opcode order*

MSD	0	1	2	3	4	5	6	7	8	9	A	B	C	D	E	F
0	BRK	ORA($aa,X)				ORA $aa	ASL $aa		PHP	ORA #dd	ASL A			ORA $aaaa	ASL $aaaa	
1	BPL	ORA($aa),Y				ORA $aa,X	ASL $aa,X		CLC	ORA $aaaa,Y				ORA $aaaa,X	ASL $aaaa,X	
2	JSR	AND($aa,X)			BIT $aa	AND $aa	ROL $aa		PLP	AND #dd	ROL A		BIT $aaaa	AND $aaaa	ROL $aaaa	
3	BMI	AND($aa),Y				AND $aa,X	ROL $aa,X		SEC	AND $aaaa,Y				AND $aaaa,X	ROL $aaaa,X	
4	RTI	EOR($aa,X)				EOR $aa	LSR $aa		PHA	EOR #dd	LSR A		JMP $aaaa	EOR $aaaa	LSR $aaaa	
5	BVC	EOR($aa),Y				EOR $aa,X	LSR $aa,X		CLI	EOR $aaaa,Y				EOR $aaaa,X	LSR $aaaa,X	
6	RTS	ADC($aa,X)				ADC $aa	ROR $aa		PLA	ADC #dd	ROR A		JMP($aaaa)	ADC $aaaa	ROR $aaaa	
7	BVS	ADC($aa),Y				ADC $aa,X			SEI	ADC $aaaa,Y				ADC $aaaa,X		
8		STA($aa,X)			STY $aa	STA $aa	STX $aa		DEY		TXA		STY $aaaa	STA $aaaa	STX $aaaa	
9	BCC	STA($aa),Y			STY $aa,X	STA $aa,X	STX $aa,Y		TYA	STA $aaaa,Y	TXS			STA $aaaa,X		
A	LDY #dd	LDA($aa,X)	LDX #dd		LDY $aa	LDA $aa	LDX $aa		TAY	LDA #dd	TAX		LDY $aaaa	LDA $aaaa	LDX $aaaa	
B	BCS	LDA($aa),Y			LDY $aa,X	LDA $aa,X	LDX $aa,Y		CLV	LDA $aaaa,Y	TSX		LDY $aaaa,X	LDA $aaaa,X	LDX $aaaa,Y	
C	CPY #dd	CMP($aa,X)			CPY $aa	CMP $aa	DEC $aa		INY	CMP #dd	DEX		CPY $aaaa	CMP $aaaa	DEC $aaaa	
D	BNE	CMP($aa),Y				CMP $aa,X	DEC $aa,X		CLD	CMP $aaaa,Y				CMP $aaaa,X	DEC $aaaa,X	
E	CPX #dd	SBC($aa,X)			CPX $aa	SBC $aa	INC $aa		INX	SBC #dd	NOP		CPX $aaaa	SBC $aaaa	INC $aaaa	
F	BEQ	SBC($aa),Y				SBC $aa,X	INC $aa,X		SED	SBC $aaaa,Y				SBC $aaaa,X	INC $aaaa,X	

Examples of use
(1) OPCODE $B0:BCS
(2) OPCODE $91:STA($aa),X
(3) All addressing modes are in the symbolic representation described in Chapter 6

Table 6.3. *Relative branch operand table*

	$0	$1	$2	$3	$4	$5	$6	$7	$8	$9	$A	$B	$C	$D	$E	$F	
$80	128	127	126	125	124	123	122	121	120	119	118	117	116	115	114	113	
$90	112	111	110	109	108	107	106	105	104	103	102	101	100	99	98	97	
$A0	96	95	94	93	92	91	90	89	88	87	86	85	84	83	82	81	Backward branch
$B0	80	79	78	77	76	75	74	73	72	71	70	69	68	67	66	65	
$C0	64	63	62	61	60	59	58	57	56	55	54	53	52	51	50	49	
$D0	48	47	46	45	44	43	42	41	40	39	38	37	36	35	34	33	
$E0	32	31	30	29	28	27	26	25	24	23	22	21	20	19	18	17	
$F0	16	15	14	13	12	11	10	9	8	7	6	5	4	3	2	1	
$00	0	1	2	3	4	5	6	7	8	9	10	11	12	13	14	15	
$10	16	17	18	19	20	21	22	23	24	25	26	27	28	29	30	31	
$20	32	33	34	35	36	37	38	39	40	41	42	43	44	45	46	47	
$30	48	49	50	51	52	53	54	55	56	57	58	59	60	61	62	63	Forward branch
$40	64	65	66	67	68	69	70	71	72	73	74	75	76	77	78	79	
$50	80	81	82	83	84	85	86	87	88	89	90	91	92	93	94	95	
$60	96	97	98	99	100	101	102	103	104	105	106	107	108	109	110	111	
$70	112	113	114	115	116	117	118	119	120	121	122	123	124	125	126	127	

Note on usage
The branch distance is a decimal number. For example,
20_{10} words forward needs branch operand $14
121_{10} words backward needs branch operand $87

6.6 Entering machine code and running it

This chapter has described the instruction set of the 6502 microprocessor and has given some short sequences of instructions that demonstrate some of the power and versatility of the simple computer's architecture. Further examples are given in Box 6.6 and a carefully graded set of exercises is given at the end of this chapter.

We must now explain how machine code can be entered into the machine memory and be made to run.

A computer cannot operate without a program. Early computers (up to third generation) were often supplied with no program in memory. It was necessary to load the first program in by hand via a switch console which would act as the bus-master. Memory locations could be addressed by switches and the contents could be read, or program code could be deposited directly. The PC could also be loaded directly and the machine could then be started. This access was needed to load and start a short program called the *bootstrap*; it gave the machine the ability to read further program code from a paper-tape reader into the computer's memory. It thus allowed the computer to 'get off the ground' by 'pulling itself up by its own bootstraps'.

Direct-access consoles are not provided on contemporary computers. Chapters 7 and 8 will show that, in normal operation, the microprocessor never surrenders mastery of the busses. This means that the machine must contain the program needed to start itself up when first turned on (*boot-up*); it will *crash* if the microprocessor encounters nonsense code. It must always be executing valid code.

We will leave further discussion of this until Chapters 7 and 8. At this point we will just describe how machine language code can be entered and run. With no direct-access console, the entering and running of machine language code must be carried out by software resident in the machine. This must be a *high-level language* program which can accept formatted information (code) entered at the computer keyboard. The resident program must interpret the code and carry out the wishes of the operator. (A high-level language is a machine language program which allows commands, programs and data to be entered into the machine in an easily learnt conversational format.) The high-level language provided for entering and running machine code in simple microcomputers is usually part of the *system monitor* or *operating system*.

Machine language access for the Apple II and BBC microcomputers are described in Box 6.5. Details for any other machine are provided in its reference or system manual.

Box 6.4
Machine language on the Apple II and BBC

When a microcomputer is in operation it must always be executing a program; when it is turned on, it must boot-up into a resident program that allows the user to enter commands and data from the keyboard. Microcomputers that must operate without a disc-based operating system usually have two resident command interpretation programs. One of these is a high-level equation-oriented language (usually a version of BASIC), the other is a simple program, the system MONITOR, which allows direct access to the memory elements. Both of these have entry points (starting addresses); if these are known it is usually a simple matter to JUMP between them. In this box we will give a brief but sufficient introduction to the Apple and BBC microcomputers to allow the reader to start writing and running machine language code.

(1) Apple II plus/IIe monitor
The machine will boot-up in the resident high-level language APPLESOFT, a dialect of MicroSoft BASIC with added commands for the graphic facilities of the Apple hardware.

The Apple also has a resident MONITOR which not only handles text input from the keyboard and text output to the display screen, but also provides a good range of commands for direct access to memory etc. We will here describe only those commands that are needed to load and run machine language code. More information is available in the relevant sections of the Apple Reference Manual.

The MONITOR starting address is $FF69, 151_{10} locations down from the top of memory (see Chapter 7).

APPLESOFT provides a command

CALL M_{10}

which causes the 6502 to execute the instruction

JSR M

CALL − 151 thus transfers the 6502 from APPLESOFT into MONITOR (the prompt changes from] to the MONITOR prompt *), with the APPLESOFT re-entry point saved on the stack.

In the following notes, *bold type* signifies user keypresses; the MONITOR response is in normal type, (CR) represents a 'carriage return'.

MONITOR expects all data and addresses in hexadecimal. Addresses are usually four digits long and data is two digits.

(*a*) *Reading memory*

The contents of any memory element can be displayed by typing its address and a 'carriage return'.

*300 (CR)

0300 – 00

*

This displays the contents of location $0300, shown here as $00.

A range of memory locations can be examined thus

*300.30F(CR)

0300 – 00 23 43 E4 00 12 DE 54

0308 – FE FD 43 65 10 FE FF 90

*

(*b*) *Writing into memory locations*

By specifying an address followed by a colon (:), the data following the colon will be deposited at the address. Successive locations can be loaded if multiple data words separated by spaces are entered before the carriage return

*300:00 01 02 1A FE

(*c*) *Loading and checking machine language code*

Machine code is loaded simply by writing the code into memory as described above. The Apple MONITOR has a very nice feature called a Dis-Assembler. This routine scans through code in memory and lists it up (one screen-full at a time) as mnemonic code, with address modes properly formatted and with the target of branches specified absolutely. The following (Example 6.12) shows these features as a MONITOR session

```
*4100: A0 0 B1 FE AA E0 0 F0 7 E0 FF F0 3 FE 0 42 C8 C0 C4 90 ED 0 (CR)

*4100L(CR)

4100- A0 00        LDY #$00
4102- B1 FE        LDA ($FE),Y
4104- AA           TAX
4105- E0 00        CPX #$00
4107- F0 07        BEQ $4110
4109- E0 FF        CPX #$FF
410B- F0 03        BEQ $4110
410D- FE 00 42     INC $4200,X
4110- C8           INY
```

```
4111- C0 C4         CPY #$C4
4113- 90 ED         BCC $4102
4115- 00            BRK
```

(d) Running a program

The 6502 can be made to run a machine language routine from MONITOR by typing the starting address followed by **G** and **(CR)**:

∗4100G(CR)

This command executes a **JSR M** instruction and therefore transfers 6502 program execution to the specified address.

If the routine ends with a BRK instruction, the 6502 will be transferred to a MONITOR routine which displays the contents of the 6502 registers and then returns to the MONITOR prompt awaiting further commands. If the routine ends with an RTS instruction, the 6502 will resume in MONITOR awaiting a further command.

A machine language routine can be called from APPLESOFT by the **CALL M₁₀** command discussed above.

If the routine ends with a BRK instruction it will end in MONITOR with the registers displayed; if ended with RTS, it will return to APPLESOFT and will resume execution at the next command.

Try these from APPLESOFT in IMMEDIATE EXECUTION mode:

POKE 768,0: CALL 768: PRINT"GONE !"

This loads $300 with BRK ($00) and then transfers execution to it.

POKE 768,96: CALL 768: PRINT"BACK !"

This loads $300 with RTS($60) and then calls it.

(e) Available memory space in the Apple

The APPLESOFT high-level language accesses practically all of the available memory space. There are simple ways to reserve some space for machine language code and to prevent Applesoft from overwriting it, but the area from $300 to $3CF is used only during boot-up and is convenient for short routines.

(2) The BBC system

The BBC computer has a different system architecture with an invisible operating system (see Chapter 8) in permanent memory. Provision has therefore been made to access memory and the PC directly from BASIC.

BASIC variables can be loaded and read in hexadecimal by use of the &

and ˜ signs:

> LET N = &FF
> PRINT ˜N

Two operator symbols (? and !) allow direct byte or word (32 bit) access to memory:

> ?&2000 = &01, sets location $2000 to $01
> PRINT ?&2000, prints the contents of $2000

> !&2000 = &A0A1A2A3, sets locations $2000–$2003 with A0, A1, A2 and A3
> PRINT !&2000, returns the contents of the four locations, A0A1A2A3

These two allow the user to write BASIC programs that will load and read memory, and will relocate blocks of code etc. The CALL &M command causes the 6502 to execute a loaded program at $M, and the RTS instruction returns the 6502 to BASIC.

The BBC BASIC includes a simple Assembler which accepts standard 6502 mnemonic code. It is engaged by the bracket operator [, and disengaged by the]. The reserved variable P% refers to the PC. The following little segment shows a very simple mnemonic code for assembly:

```
100 P%=&2000
110 [
130 LDA #&0F
140 TAX
150 STX &2008
160 RTS
170 ]
```

When this program is RUN, it assembles and loads the 6502 machine code starting at $2000 and outputs a listing of the mnemonics and the machine code onto the screen.

Care must be taken to ensure that the BASIC code in the machine and the display screen are kept clear of the machine code: $2000 is fairly safe with short BASIC code and MODE 7 text display. Page $0D ($0D00–FF) is available if the BBC has no disc or network interfaces installed. (For more information, see the references at the end of this chapter.)

(3) Argument transfers between BASIC and machine language routines

BASIC in the Apple and the BBC has two commands, PEEK and POKE, which allow data to be read or written into memory locations. These two allow data to be passed to, or recovered from, machine language routines. Both versions of BASIC also define a **USR** function which provides a limited variable pass in both directions. The BBC reserves the variables A%, X%, Y% and C% for the A, X, Y Registers and the C bit in USR transfers.

It is possible to access BASIC variables directly from machine language routines, but this requires a detailed knowledge of the variable structure within the dialect of the machine's BASIC language. There is insufficient space to go into this here, the interested reader is referred to publications from the computer user's groups. Some discussion of this for instrumentation applications is made in Chapter 8.

6.7   Some final remarks

The provision of a minimal but complete instruction set in an 8-bit microcomputer does not seriously restrict the range of tasks that it can carry out, but does effect the speed and length of the machine code. With the more extensive instruction set of 16-bit and larger machines, the programmer can usually find several ways of executing a task; the program can be optimised for short or fast code. In a small 8-bit machine, the problem is often to find *any* way to implement a particular algorithm within the minimal instruction set. This is explored in the examples in Box 6.6 and in the exercises at the end of this chapter. Some references for further reading are given at the end of this chapter.

The simple instruction set of the 6502 allows machine language programs to be written directly in machine code. This is only efficient, however, up to about a page or so of code. Beyond this size, the programmer should be using an Editor/Assembler program to generate machine code. Microprocessors with complicated instruction sets (and particularly those with abundant and complex addressing modes) are not suited to direct machine code programming; the coding of the instruction and its addressing mode requires selective setting or clearing of individual bits of the instruction code-word. This is quite tricky and is very prone to human error; a computer program is much better at this type of task and an assembler is almost mandatory.

An Editor/Assembler program is in two parts. One of these is a high-level language text editor program that allows code to be written and edited in the microprocessor's standard mnemonic operation and addressing mode formats. Memory locations and instruction codes can be labelled with names and used like variables within the code. (Some of the examples in this chapter have been written in Assembler format to make it look familiar to you.) This allows the algorithm code to be *position independent*. Actual values for variables, labels and locations are assigned in an equate or assignment table; modifications and reassignments are easily effected.

The second part of the Editor/Assembler uses the text generated by the Editor as the *source code*. It reads the code at least twice. The first pass tabulates all of the labels, etc., to evaluate addressing mode (and branch) operands; the second pass is used to convert the source code into the final machine language code, the *object code*.

Some assemblers allow totally relocatable code to be written even when absolute jumps etc. are used. These store a listing of all addresses that must be relocated and consult this when the program is finally loaded into memory by a *relocating loader* program.

Most Assemblers allow several source code text files to be linked together at assembly time. This allows large programs to be created from a library of standardised, short and easily *debugged* routines.

Many assemblers allow *macroassembly*, where predefined code segments (*macros*) can be invoked by name in the source code and included automatically (and even conditionally) into the object code at assembly time.

Box 6.5
Machine language examples

Example 6.13

The 6502 provides only one stack. It is possible, however, to *partition* the stack. The code below depicts the generation of a private stack partition for use therein. (See also Exercise 6.14.)

Code and notes

```
TSX              :Transfer the stack at subroutine entry
STX ESS          :to a reserved location and then load the previous
LDX ISS          :private stack pointer
TXS              :into the Stack Pointer
 ::  ::
 ::  ::
TSX              :At the end of the routine,
STX ISS          :reverse the procedure, saving the private
LDX ESS          :stack and restoring the previous
TXS              :stack pointer
RTS
```

Example 6.14

(*a*) This routine provides a one-page software stack starting at the location defined by STAKU. The two entry points allow the Accumulator contents to be either pulled or pushed onto this pseudo-stack.

Code and notes

```
PUSH STX XSAVE        :Save X register
     LDX STPTR        :Pickup stack pointer
     STA STAKU,X      :Save accumulator
     INX              :and point to next element
     STX STPTR        :Save new stack pointer
     LDX XSAVE        :and restore old X
     RTS
PULL STX XSAVE        :
     LDX STPTR
     DEX              :decrement pointer
     LDA STAKU,X      :before fetching data
     STX STPTR
     LDX XSAVE
     RTS
```

(*b*) The restricted indirect addressing modes provided in the 6502 preclude neat user stack routines. The segment below is an example of *self-modifying code*. Such code can provide very compact and fast routines under carefully controlled conditions.

The routine shown cannot be position independent; it is shown placed in general memory space starting at $40F0, but it would execute much faster if sited in PAGE #0. It saves the accumulator contents in a stack extending indefinitely upwards from location $4100 (PSTACK). The reader should attempt the complementary routine to PULL data off this pseudo-stack.

Code and notes

```
PUSH STA PSTACK     STA $aaaa    $40F0: 8D 00 41
     INC STPTRLO    INC $40F1           EE F1 40
     BNE END        BNE $03             D0 03
     INC STPTRHI    INC $40F2           EE F2 40
END  RTS            RTS                 60
```

Example 6.15

The 6502 provides no instructions that access the PC directly. The subroutine below returns, in the A and X Registers respectively, the low and high bytes of the instruction opcode which follows the subroutine call.

Code and notes

```
PLA        :The last word on the stack
TAY        :is the return address low byte
PLA        :The second last is the high byte
TAX        :which is put into X
PHA        :and then restored to the stack
TYA        :The low byte is then put into A
PHA        :and restored (stack is now as it was)
RTS
```

Example 6.16

(Refer to Section 3.4 and to Box 3.5.) The instruction set of the 6502 allows simple implementation of the algorithm for long multiplication. In this example, a simple routine to carry out long multiplication between two 8-bit numbers is shown. Recall that such a multiplication will have a result of up to 16 bits; thus a 16-bit accumulating register will be needed. First, let us set up memory locations for all the variables used.

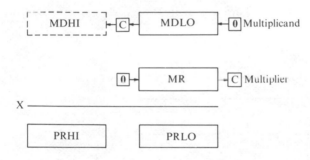

The two variables MDLO and MR are the multiplicand and the multiplier respectively; they will be loaded before the routine is carried out. Successive multiplication of the multiplicand by two is carried out by left-shifts, with the m.s.b. passing into the MDHI location. At the start of the routine, this location and the two-word product register must be cleared to zero.

The long multiplication algorithm requires that the multiplier word be scanned bit-by-bit as the multiplicand is successively shifted; if the relevant multiplier bit is true, the present multiplicand is added into the product. The instruction set of the 6502 carries this out most easily if the multiplier is shifted right rather than scanned bit-by-bit; the bit 'falling out' can be easily tested. The procedure will be finished when the residue in the multiplier reaches zero.

Code, notes and flow chart

```
            LDA  #$00      :(A)
            STA  MDHI
            STA  PRHI
            STA  PRLO
      TEST  LSR  MR        :(B) l.s.b.→ C
            PHP            :(C) Save Zero flag
            BCC  MULT      :(D)
            CLC            :(E)
            LDA  PRLO
            ADC  MDLO
            STA  PRLO
            LDA  PRHI
            ADC  MDHI
            STA  PRHI
      MULT  ASL  MDLO      :(F)
            ROL  MDHI
            PLP            :(G) Recovers Zero flag
            BNE  TEST
            RTS
```

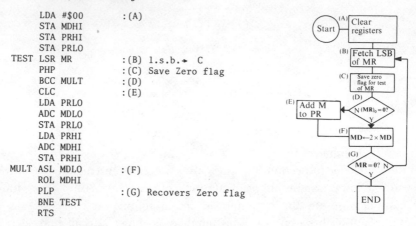

The algorithm below is another version which works in the reverse order, with the multiplicand going into MDHI. Note that the multiplicand must be halved at the start of the multiplication process.

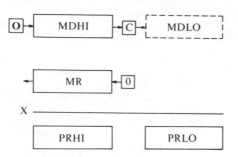

Code, notes and flow chart

```
            LDA  $#00
            STA  MDLO
            STA  PRHI
            STA  PRLO
            LSR  MDHI       :Halve the MD to adjust
            ROR  MDLO       :the bit weighting
            BIT  MR         :First MR m.s.b. fetch
      TEST  BPL  DIVD
            CLC
            LDA  PRLO
            ADC  MDLO
            STA  PRLO
            LDA  PRHI
            ADC  MDHI
            STA  PRHI
      DIVD  ASL  MDHI
            ROL  MDLO
            LSR  MR         :Subsequent MR m.s.b. fetch
            BNE  TEST
            RTS
```

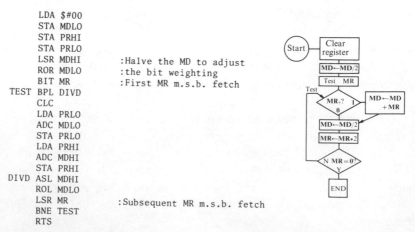

Multiplication can of course be carried out by repeated addition of the (fixed) multiplicand into the product for the number of times specified by the multiplier. This is easy to program but is very slow in execution.

Example 6.17

(Refer to Section 3.5.) The algorithm for long division of binary numbers is also easy to implement in machine language. It is simply the inverse of the algorithm above, but compact, fast coding requires a few tricks. The algorithm presented below requires that the initial dividend (DDLO) be moved bit-by-bit into another word (DDHI). At each move, the divisor (DR) is compared with DDHI. If DDHI is less than DR, the corresponding quotient bit should be 0, if DDHI is greater than or equal to DR, the quotient bit must be 1 and the DR contents must be subtracted from DDHI. Note how the use of the Accumulator speeds up and shortens the code. The quotient is generated bit-by-bit, by eight successive left shifts from the C bit; the result of the DDHI, DR comparison should thus be manoeuvred into the C bit. Behold!

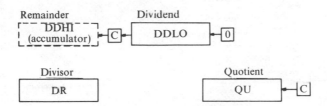

Code, notes and flow chart

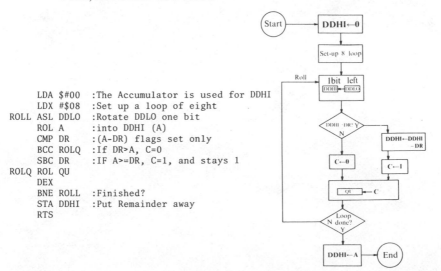

```
          LDA $#00   :The Accumulator is used for DDHI
          LDX #$08   :Set up a loop of eight
ROLL ASL DDLO        :Rotate DDLO one bit
     ROL A           :into DDHI (A)
     CMP DR          :(A-DR) flags set only
     BCC ROLQ        :If DR>A, C=0
     SBC DR          :IF A>=DR, C=1, and stays 1
ROLQ ROL QU
     DEX
     BNE ROLL        :Finished?
     STA DDHI        :Put Remainder away
     RTS
```

Example 6.18

Floating-point representation of numbers was described in Section 3.6. A simplified model was presented in Box 3.9. The routine below performs binary multiplication between the two 8-bit mantissae of numbers in this model and then adjusts the exponents to complete the floating-point multiplication. The routine is not intended to be fast or compact; it is written in a form that makes it easy to follow. Note the tests performed at the beginning to handle zero arguments, and the tests carried out at the end to handle exponent over/underflows.

Flow chart, notes and code

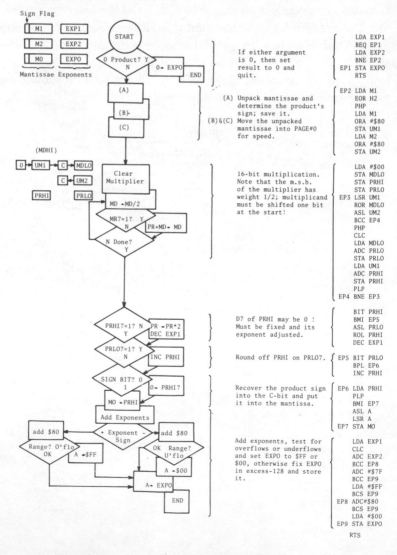

Exercises

6.1 Rewrite the routines of Examples 6.1 and 6.2 (Box 6.2(*a*)) such that the transfers are made via ZERO PAGE addresses. Compare the number of cycles and bytes required for both techniques. (Use Table 6.1.)

6.2 Devise an algorithm and write a routine to swap the contents of the A and X Registers of the 6502. Attempt to use the STACK, but if this is too difficult use the technique of Example 6.5 (Box 6.2(*b*)). Calculate the cycle count for your routine.

6.3 Determine the full bus transaction sequence that must be generated by the instruction decode logic of the 6502 to execute the LDA operation in all of its addressing modes. Compare your required cycle count with that actually required (see Table 6.1). Can the indirect X and Y modes be carried out with only the registers described in Chapter 5?

    You might also like to go through the bus timing diagrams in reference 3 at the end of this chapter, to derive how the IDL uses the busses during instruction executions.

6.4 Determine the control signal sequence and register usage to execute the instruction

JSR ($aaaa)

6.5 (Important.) Show that the three operations **AND, ORA** and **EOR** allow any bit of any word to be cleared, set or negated. Write a routine to demonstrate this. Operate your machine language routine from BASIC by POKEs, PEEKs and CALLs; assemble binary keyboard entries consisting of strings of **1** and **0** into binary words, pass them to your routine and unpack the result into a multi-bit binary display.

6.6 Attempt to devise a more compact or faster routine to carry out an 8-bit rotation (see Example 6.8 (Box 6.2(*c*))).

6.7 Consider Example 6.9 (Box 6.2(*d*)). What changes (if any) are needed to make the routine function properly for signed integer (two's-complement) numbers?

    Add code to watch for numerical over/underflow and to signal these by some means to the driving program. (Use a BASIC driver program to load the variables and return the results.)

6.8 Modify the routine of Exercise 6.7 to work with two-word (16-bit) two's-complement numbers. Rearrange it so that there are two entry points; one adding the two numbers, the other subtracting them. Provide over/underflow warnings.

6.9 (*a*) Modify the code of Example 6.12 (Box 6.2(*e*)) so that the Y Register is first loaded from a ZERO PAGE memory location and the loop continues until the Y Register reaches $00.

    (*b*) Extend the code so that the starting address and length of the scanned table can be defined by two-word (16-bit) variables freely accessible to a calling routine elsewhere in memory.

(*c*) Add a test within the routine to terminate it when any element of the tally table reaches $FF. (Test for a result of $00, then decrement it and exit.)

6.10 Write a routine to add together two arrays each consisting of 128_{10} two-word numbers, putting the result in the first array. Use all suitable addressing modes and calculate the speed/space trade-off for each mode.

6.11 (*a*) Modify and extend the routine of Example 6.16 (Box 6.6), or devise a different routine to handle two-byte (16-bit) multiplication.
(*b*) Do the same for 16-bit division (Example 6.17 (Box 6.6)).
(*c*) Extend both routines to handle unsigned two's-complement integers.

6.12 Modify or rewrite the routine of Example 6.18 (Box 6.6) to perform (*a*) floating-point division, (*b*) floating-point addition. Ensure that your routines handle signed magnitudes and zero arguments properly. Note that in (*a*), the first comparison of the division algorithm yields a quotient of 1, not 0.1 (1/2).

6.13 Devise an algorithm and write a routine to scan through a predefined block of memory (starting address and length specified as variables), searching for a predefined *string* of predefined length. Make it tabulate an array of pointers to all occurrences of the string.

6.14 Extend the routine of Example 6.13 (Box 6.6) to watch for interference between the two partitioned stacks. Return a specific code in the Accumulator to advise what has happened.

6.15 Use the concept presented in Example 6.15 (Box 6.6) to write a routine which, by executing a JSR to an RTS instruction, determines its own address in memory.

6.16 Read the references in Chapters 1 and 6 and other sources about the Analytic Engine of Babbage and the Turing machine. Write an essay on the conceptual differences and similarities between the Babbage/von Neumann architecture and the Turing architecture.

References and further reading

This list is a brief resumé of some of the many books and articles available.

1 'Turing and the Origin of the Computer', *New Scientist*, 21 May 1984, pp. 580–5
2 'Turing Machines', *Scientific American*, May 1984
3 *The SY6500 Microprocessor Family – Applications Information AN2*, Synertek (USA), 1980
   This is a hardware-oriented applications note for the 6502.
4 Reference manuals for the Z80, 6809, 6502C etc. microprocessors
5 *The 8086 Primer*, S. L. Morse, Hayden, New York (1980)

6 *6502 Machine Code for Beginners*, A. P. Stephenson, Newnes, London (1983)

7 *Assembly Language Programming for the Apple II*, R. Mottola, Osborne/McGraw-Hill, Berkeley (1982)

8 *6502 Assembly Language Routines*, L. A. Leventhal & W. Saville, Osborne/McGraw-Hill, Berkeley (1982)

9 *Assembly Language Programming on the BBC Micro*, J. Ferguson & T. Shaw, Addison-Wesley, London (1983)

10 'Applesoft Internal Entry Points', *Apple Orchard*, March/April 1980 (the magazine of the Apple Users Group)
   This describes many important routines within Applesoft and shows how to access them from ML routines.

11 *Apple 6502 Assembler/Editor*, Apple Inc. (USA) (1980)
   This easy-to-use Assembler, with a relocating loader, is part of the 'AppleDOS Toolkit' software enhancement package for the Apple II.

11 *MACRO-86*, Microsoft Inc. (USA)
   This is a rather powerful macro- and conditional-assembler for the 8086; it is part of the MS–DOS system for 8086/88-based microcomputers.

12 *Discover FORTH*, T. Hogan, Osborne/McGraw-Hill, Berkeley (1982)

7

The computer/world interface

Preamble

The three previous chapters have described how the architecture of a computer provides the ability to perform any task that can be represented as a sequence of logical operations on a data set held within the machine. The ability to explore logical and mathematical models of natural processes makes the computer particularly useful in the fields of Science and Engineering (Section 1.1). (See the reading list at the end of this chapter.)

There is, however, a gross difference between the computer's internal model and the natural process in the outside or 'real' world. The computer model is a *sequentially executed program* of discrete logic operations on *binary logic variables*. It can only manipulate one variable at a time, whereas the outside world has multiple, continuously varying and simultaneously interacting parameters.

No problems arise when a computer is used to perform *data-processing* or numerical analysis on a prepared database, provided, of course, that the program and its database fit into the limited memory of the machine and that it finishes execution while the results are still relevant. The only limits to the application of the computer are its size and speed. However, if the computer must interact directly with the outside world (when the computer program must run in *real-time*), the gross difference between the program model and the outside world requires careful consideration.

This chapter will deal with the two major problems of real-time microcomputer applications. One of these is the electronic logic connection between the computer and the outside world, the computer hardware *interface*. The second, harder, problem is in ensuring that the two disparate systems can establish and maintain satisfactory *interaction*.

The principles of interface logic circuitry will be described in sufficient detail to support the design of simple interface circuits. For continuity with previous chapters, and because of the open, friendly architecture of the Apple, the principles will be demonstrated within that machine. An outline of extension to the BBC will also be given. The same principles apply to all other microprocessors and microcomputers. In fact, the fully timed control signals of the 8080, Z80, 8086/8, 6800, 68000, etc., make the design of their interface hardware a little easier. The interested reader should be able to apply the principles to any other microcomputer, with the aid of its hardware reference manual.

More examples of interfaces and interaction schemes will be described in Chapter 9.

7.1   Computer/world interaction

When a computer is used in a real-time control or data-collection application, information must usually flow in both directions between the computer program and the outside world. Primary transducers sense relevant outside-world variables and convert them to electrical signals. (A discussion of transducers is outside the scope of this book; the reader is referred to the literature.) The transducer outputs must be converted into binary logic signals suitable for data input to the computer program. Thus, the computer program can monitor relevant parameters in the outside world and process this information. If it is to control some aspect of the outside world, it must decide how changes can be effected, and it must generate output *control* signals through the interface to the outside/real-world system, via control transducers.

The commonest example of this is where the outside world is just the programmer or operator. Input to the system is the data entered at the keyboard or from disc etc. as required by the program. The computer output is presented to the outside world on the display (VDU), printer etc. as soon as the program has it available. In this situation, the computer operates at its own speed and the outside world does its best to keep up or not to get bored. The program has control and determines when the input device (the operator) must present data. It issues control commands via the display to which the input device must respond. In this example, the execution of the program is usually so much faster than the response time of the real-world element that no interaction problems occur. The disparity between the two systems is hidden by the speed of the computer and the slow, serial data I/O.

A similar example is where a computer is used to make measurements on an external system. The computer program is in full control of the sequence and determines the rate at which control parameters are adjusted and measurements are made. In both of these examples, the program is the system *master* and the outside-world system is a *slave*. Provided that the data-collection rate is within the capability of the computer program, there are no interaction problems.

The more difficult cases are where the computer program is not in control and must respond as *slave* to changes in parameters of the external, outside-world *master* system. The outside system may have several parameters changing simultaneously, all demanding prompt computer response. Since the computer is strictly sequential, it cannot respond simultaneously to all demands; it must *service* them in sequence. If the external system changes very slowly, there will be no serious problem; however, it if changes quickly, the computer may not be able to 'keep up' in 'real-time'.

We will consider all of these interaction situations in this chapter, but we will consider first the *interface hardware* between the outside world and the computer.

7.2   The computer/world hardware interface

The content of Chapters 4, 5 and 6 indicates that the outside world can only interface to a computer by data transactions on the computer databus. There are two obvious techniques for this.

(1) One technique is for the outside world to simply write or read data to or from the bus system through normal slave elements (usually called *I/O ports*) in the computer's address space. These are exactly like normal data transactions and must be controlled by the processor and its program.

(2) The second technique is for I/O data transactions to be carried out directly with the computer's address space by an alternative bus-master element. This is usually called *Direct Memory Access (DMA)*.

We will deal first with the prior technique, but we will return later to discuss the second technique and some of its subtle derivatives.

The notion of an I/O port was discussed in Chapter 4. The I/O port logic circuitry uses the address, control and data busses in the usual way, as described there. Box 7.1 describes a simple digital logic circuit for an 8-bit bidirectional I/O port into the Apple microcomputer. The provision made in the Apple and the BBC for I/O ports is also described.

A wide range of special-purpose VLSI interface chips is available for all microprocessors, but their black-box structure obscures the essential simplicity of interface hardware. To provide full disclosure, we will therefore describe interface circuits in this chapter, built mainly from simple TTL logic gates. The special chips will be described and evaluated in Chapter 8.

Box 7.1
A simple digital interface

The 6502 is a memory-mapped I/O microprocessor; the Apple microcomputer must therefore set aside some of its address space for I/O ports. The memory map of the Apple is just like that shown in Fig. 4.8 (Box 4.2). The top quadrant of memory is used for the I/O and the resident system *firmware*. (Firmware is the name given to the programs which are held in *ROM*; this is discussed later in this chapter and in Chapter 8.)

Fig. 7.1 depicts the assignments. The region from $D000 to $FFFF is allocated to the resident firmware consisting of 2K of MONITOR code and 10K of APPLESOFT code. Addresses from $C000 to $CFFF are reserved for I/O and general peripheral use. The Apple confines all of its inbuilt I/O ports (keyboard, screen, tape etc.) to the region from $C000 to $C07F. The remaining region from $C080 to $CFFF is allocated and decoded for

Figure 7.1   The Apple ROM and I/O mapping.

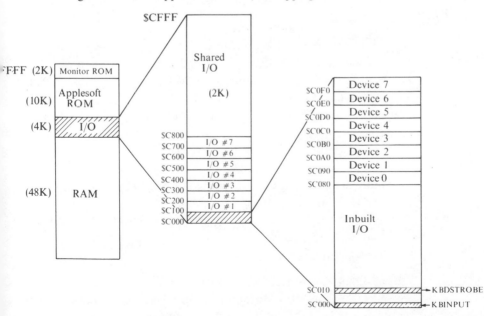

external device interfaces. The provision of this space, and the ease of access to it, has made the Apple the DC3 of personal computers.

The Apple II has eight (the Apple IIe has seven) *peripheral slots* for printed circuit *edge-connector* boards. Each connector provides all of the computer's databus, address bus and control bus signals as well as some address decode signals described below. The connectors are named SLOT #0–SLOT #7.

Address decode logic within the Apple main circuit board provides decoded signals corresponding to each of the subregions depicted in Fig. 7.1.

(1) When the address bus holds an address between $C800 and $CFFF, that is when the address pattern is

   1100XXXXXXXXXXXX

   (where **X** signifies either **1** or **0**), a control signal named *I/O STROBE* goes low during the data phase of the bus transaction cycle. This signal is presented at all slots and indicates that the bus-master is accessing the 2K block of memory set aside for shared use of devices in any slot. A protocol is defined in the Apple reference manuals to allow multiple peripherals to share this area. It is intended for software to service extra peripheral devices, but it can also be put to good use by instrumentation peripheral hardware.

(2) When the address bus holds a pattern between $C100 and $C1FF, that is,

   11000001XXXXXXXX

   a signal called I/O SELECT # 1 and presented only to SLOT # 1 goes low during the data phase of the bus transaction cycle. Since this signal is only presented to SLOT # 1, it assigns this page of addresses to that slot. It is intended principally for short program code carried in memory on the peripheral device plugged into the slot. A peripheral should only use the page assigned to its slot.

   Similar address decode signals assign contiguous pages individually to SLOT #2 through to SLOT #7 (see Fig. 7.1). SLOT #0 does not have a page assigned.

(3) When the address bus holds a pattern between $C080 and $C08F, that is,

   110000001000XXXX

   a signal called DEVICE SELECT #0 and presented only to SLOT #0 goes low during the data phase (the Apple IIe has no

SLOT #0). This signal assigns these 16 addresses to that slot. This small region is intended as I/O port addresses for the peripheral in this slot.

Similar decode signals are provided at all other slots (see Fig. 7.1).

These signals simplify the decoding logic needed in external peripheral circuitry. The combination of ample I/O space for external devices, address decode signals, a transaction timing signal Q_3, and simple, open architecture, makes the Apple a very simple and convenient machine to interface.

Fig. 7.2(*a*) shows a simple, minimal circuit to provide an 8-bit digital I/O interface port using the signals provided at an Apple peripheral slot. Fig. 7.2(*b*) depicts the timing for data transactions in either direction. This should be studied in conjunction with Figs. 4.17 and 4.18 in Box 4.4.

Note the following features of Fig. 7.2.

(1) Since only one of the 16 assigned addresses is needed, decoding is simplified by allowing the bus driver/receiver chips to respond to transactions with all 16 DEVICE SELECT addresses.

(2) The data at the digital input is latched into chip 2A before the latched data is deposited on the databus. This ensures that the data is stable during the transaction.

(3) The simple gating performs full timing for R/W operations. (The Q_3 signal makes a 1/0 transition when the 6502, writing, would have stable data written on the bus.)

Figure 7.2 (*a*) Eight-bit digital I/O port, (*b*) data transaction timing diagram.

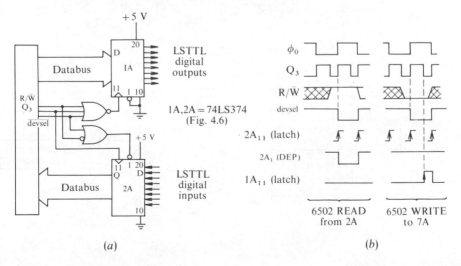

(*a*) (*b*)

The simple circuit of Fig. 7.2 squanders all the addresses assigned to the DEVICE SELECT signal on one 8-bit I/O. The circuit of Fig. 7.3 allows all 16 assigned addresses to be separated (decoded), to provide 16 separate R/W functions in the interface. The 74LS138 3–8 line decoder chip is used; its truth table is shown in Table 7.1. (The 74LS155 and the 74LS139 decoders chips should also be studied.) The 74LS138 chip has three control inputs, all of which must be in their active state for the logic outputs to be enabled. (These decoders can therefore be chained into a *decode tree* by connecting higher-order decode outputs to lower-order decode enable inputs.)

The enable inputs provide two functions, to synthesise a 4–16 line decoder from two simpler chips, and to provide full transaction timing for both reading and writing. The two upper decoders 3B and 4B provide 16 separate, active-low OUTPUT ENABLE signals timed, as in Fig. 7.2, to suit 74LS374 bus drivers. Full 4-bit (16 address) decoding is achieved by driving chip 3B directly from the address bus line A_3 into its G_3 control input (active-low) so that it will respond only to the first 8 DEVICE addresses assigned to the slot. The negation of A_3 is fed to G_3 of 4B to respond to the second 8 assigned addresses. The R/$\overline{\text{W}}$ line from the control

Figure 7.3   (a) Address decoding logic, (b) R/W timing and (c) a bidirectional databus buffer.

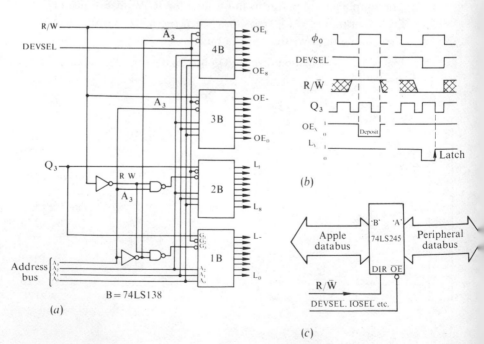

(a)

(b)

(c)

bus is presented to both 3B and 4B at the G_1 enable input so that these chips only respond while the processor is *reading* from the databus.

The two decoders 1B and 2B are addressed like 3B and 4B, but the enable inputs are reorganised to achieve enabling of the chip outputs only when the processor is *writing* to the databus, and to time the positive-going latch edge while the 6502 has stable data on the databus. Fig. 7.3(b) depicts the timing for both signals.

These signals can be used for any operation such as setting or clearing flip-flops, counters etc. Simpler timing can be used for these operations, but note well that while the 6502 IDL is interpreting some addressing modes, the address bus may hold improper addresses; simple decode logic may generate spurious signals. The R/$\overline{\text{W}}$ line will always be high during these false addresses, hence it is advisable to generate the decode signals (usually called *strobes*) from 6502 write instructions. (Refer to the 6502 hardware manuals for more information.)

The address and data busses of microcomputers are usually heavily loaded and cannot tolerate a lot of extra loading from peripherals. If many devices must be added, or if there is any chance of the peripherals abusing the computer busses, it is advisable to *buffer* them from the computer busses. Address busses can usually be buffered with unidirectional chips such as the 74LS244. A bidirectional databus buffer can use the 74LS245 bus transceiver (Fig. 7.3(c)).

If care is taken with the decode logic timing in the interface, the slight data delay introduced by the databus buffer can allow the end of the data

Table 7.1. *Truth table 74LS138 3–8 line decoder*

Enables		Inputs			Outputs							
G_1	G_2+G_3	A_0	A_1	A_2	Q_0	Q_1	Q_2	Q_3	Q_4	Q_5	Q_6	Q_7
X	1	X	X	X	1	1	1	1	1	1	1	1
0	X	X	X	X	1	1	1	1	1	1	1	1
1	0	0	0	0	0	1	1	1	1	1	1	1
1	0	1	0	0	1	0	1	1	1.	1	1	1
1	0	0	1	0	1	1	0	1	1	1	1	1
1	0	1	1	0	1	1	1	0	1	1	1	1
1	0	0	0	1	1	1	1	1	0	1	1	1
1	0	1	0	1	1	1	1	1	1	0	1	1
1	0	0	1	1	1	1	1	1	1	1	0	1
1	0	1	1	1	1	1	1	1	1	1	1	0

phase to be used to latch data from the buffered databus. This simplifies the data–latch timing circuit.

A pair of I/O ports can provide I/O port or memory expansion. If one output port is regarded as a supplementary address register, it can be used to expand the other port to 256 separate ports. Access is gained to any one of these supplementary ports by first depositing its address in the register port, then communicating with the expansion port via the second I/O port. This maps a full page of addresses through a pair of actual bus addresses (Fig. 7.4(*a*)).

   The logical extension of this concept allows one output register and one page of I/O addresses to provide access to 64K of expansion memory or I/O space (Fig. 7.4(*b*)). Further expansion can be made by providing a second page pointer to define 64K pages!

BBC interfacing

   The BBC microcomputer interface facilities are more restricted. The bus hardware can only be accessed via the 1 MHz expansion bus socket or the Acorn TUBE. The 1 MHz expansion socket provides buffered outputs from the bottom eight lines of the address bus. Buffered bidirectional access to the eight databus lines is enabled when the processor is accessing addresses $FC00–$FDFF. These two pages, the third and fourth last pages in memory, named Fred and Jim, are not used by the internal architecture. They are within the system ROM, but decode logic disables the ROM when these addresses are accessed. Two signals, PGFD or PGJM, indicate when the 6502 is accessing these pages. (The address expansion technique of Fig. 7.4 can be used for enhancement.) The 2 MHz system clock is dropped to 1 MHz during transactions. A 1 MHz system

Figure 7.4   (*a*) Port expansion scheme and (*b*) memory expansion by page overlay.

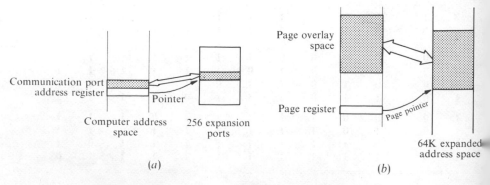

(a)   *(b)*

Figure 7.5 (a) BBC memory I/O map, (b) the 1-MHz interface and (c) the Acorn TUBE.

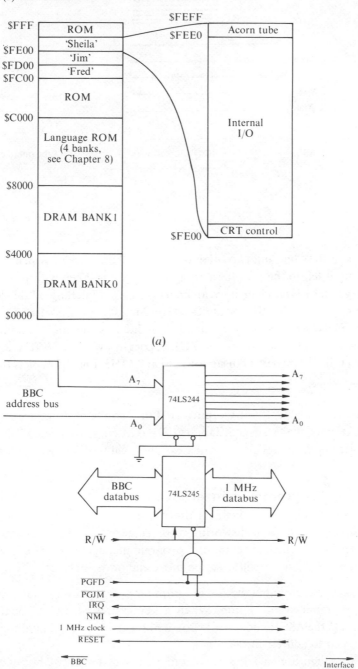

(a)

(b)

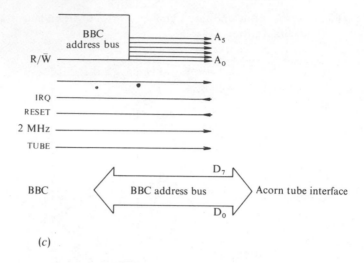

(c)

clock, a R/$\overline{\text{W}}$ line and two other 6502 control lines are provided. Detailed timing is left to the peripheral designer; if care is taken with propagation delays, the 1 MHz clock line can be used for data latching off the databus.

The Acorn TUBE occupies 32 addresses ($FE00–FEFF); direct unbuffered access is provided for all eight databus lines and the address bus lines A_0–A_5. The decode signal, TUBE, goes low when the 6502 is accessing the TUBE addresses. Transactions are at 2 MHz. Fig. 7.5 depicts the logic.

Programmable Input/Outputs (PIO), Peripheral Interface Adapters (PIA), Versatile Interface Adapters (VIA) etc. chips are available for direct interfacing to microcomputer busses. They will be discussed in Chapter 8.

7.3   Programmed interaction

This section considers the problem of interaction between the computer and the outside world. At this stage only the simplest scheme will be described, where the computer program manages the entire interaction process. Section 7.6 will describe more elaborate schemes.

The keyboard input of the Apple computer provides a very simple example of programmed interaction. When a key is pressed, the keyboard logic presents the ASCII code word (Box 2.4) for the key to a digital input port assigned to address $C000.

Since there are only 96 symbol and control codes, only 7 bits of data are needed. The unused bit, D_7 (normally the parity bit), is used as a *status* flag.

When a key is pressed, the keyboard logic also sets a flip-flop connected to this bit. The keyboard logic gives no other signal, thus any program needing keyboard input must read from this port to test the m.s.b. status and to determine when a key has been pressed. The status flip-flop can be cleared by addressing another location ($C010), called the CLEAR KEYBOARD STROBE. (This is not a data I/O address, accessing only *strobes* the keyboard status flag clear as discussed in Box 7.1.)

To detect a keypress, the program must repeatedly test the input port until the status flag goes high

```
LDA $C000    :Read keyboard input
BPL $FB      :Loop back unless status high
STA $C010    :Clear keyboard status flag
OPC  XXX     :Response program
```

The program segment above loops until a key is pressed, when it clears the status flag and moves on to the response.

This type of program-driven interaction with the outside world requires no hardware other than an input port, and is thus very easy to implement, but it is not suitable for all applications. Prompt response can only be achieved by total program commitment to watching the status flag; this prevents the computer from carrying out any other tasks.

An alternative scheme is for the program to test the status periodically between other tasks; this, however, does not guarantee prompt response to outside-world events. This scheme is called *status polling*. APPLESOFT BASIC provides a simple example, associated again with keyboard input. When BASIC is running (interpreting) a user program, it tests the keyboard input after executing every BASIC command. If the keyboard input port holds the code for a control-C ($83) keypress, BASIC program execution is terminated and the machine returns to COMMAND (input) mode.

This polled, program-driven scheme is suitable for interaction with the outside world where moderate response times are adequate.

The more general case for control or instrumentation applications requires several inputs with independent status flags to be monitored. These status signals can be monitored by *polling* all relevant input ports to allow the program to decide what actions are needed, and the order or *priority* of the actions required.

If multiple, independent, peripheral interfaces are used, each must have its own status register or status flag. If a composite interface with many data inputs is built as a single peripheral system, it is preferable to combine all the status flags into one or more words at a dedicated status input port.

A single data read from this port can then determine the entire status condition.

Some simple examples of these techniques are described in Box 7.2.

Box 7.2
Simple status polling

In these examples we will assume data I/O hardware as described in Box 7.1 and consider only the interactions between the program and the STATUS flags at the input ports.

Example 7.1

A simple subroutine to poll the status of several peripheral ports, each flagged at the m.s.b. of its status word or input port, and to branch to required service routines.

ENTRY	LDA	STATUS0	: Fetch first status word
	BPL	TEST1	: If flag low, go to next
	JSR	SERVICE0	: If flag high, service it
TEST1	LDA	STATUS1	: Fetch next status
	BPL	TEST2	: Etc.
	JSR	SERVICE1	:
TEST2	::	::	:
	::	::	:
	RTS		: Finally, return

The order in which the flags are serviced is determined by the test sequence in the program. A high-priority flag which arrives after the subroutine has polled its status, will not be serviced until all lower-priority services have been completed and the poll routine is reentered.

Example 7.2

This service routine shows how a single status word holding several status flags can be analysed quickly to test for a null service condition.

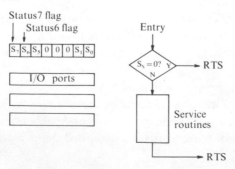

```
ENTRY        LDA   STATUS      :Fetch composite status word
             BEQ   END         :If no flags, return
             JMP   SERVICE     :If any flag, service
END          RTS               :
SERVICE      ASL               :Test m.s.b. of status word
             BCC   NEXT1       :If low, try the next bit
             PHA               :Save A for next test
             JSR   SERVICE7    :If high, service it
             PLA               :Recover status word
NEXT1        ASL               :
             BCC   NEXT2       :
             ::                :
             ::                :
             RTS               :Back to main program
```

The priority of service is determined by the order of flags in the status word. Once the status word is read, all flags will be serviced in strict priority order before the next status read.

Simple modification of the routine above allows only the highest priority flag to be serviced at each poll of the status word. This provides fast response to urgent status changes.

Simple programmed interaction can be implemented with a minimum of hardware, and is quite adequate when the response of the program to outside events need only be of the order of milliseconds. Status polling via a single status word can be achieved in about 20 machine cycles. The response time is determined by the execution time of the service routine and by the frequency of status polling.

7.4   Analog variables and their conversion

The term *analog variable* is commonly used to distinguish a continuously variable electrical signal from binary logic variables. The name comes from the *analog computers* which were used extensively, before the digital electronic revolution, to analyse the behaviour of complicated dynamic systems. They contained arrays of *operational amplifiers* which could be configured to model the mathematical operations of scaling, summation, differentiation or integration on time-varying input voltages (see references). By connecting these operations as a network an analog computer could model linear integro-differential equations.

Any dynamic system which could be represented by such equations could therefore be modelled, by its *electrical analogue*, on the computer. Its dynamic behaviour under any conditions could then be examined by varying the driving function (the time-varying input voltage); its behaviour could be displayed on a *Cathode Ray Oscilloscope (CRO)*. The real system could thus be analysed, and its sensitivity to manufacturing tolerances in its components could be evaluated by varying the analogue model. This was very convenient, but these analog computers were limited to a few per cent accuracy. Digital computers now have sufficient capacity and speed to perform this modelling to any required precision, within the convenience of software. They have therefore replaced the analog computer, leaving only half of its name in zippy spelling, to refer to the use of electrical signals to represent continuous real-world variables.

Measurements of continuously variable (analog) parameters must still be made, and many computer-controlled systems need continuously variable output signals. Convertors are needed, therefore, to provide *Analog to Digital Conversion (ADC)* from the outside world into the computer, or *Digital to Analog Conversion (DAC)* from the computer to the outside world.

7.4.1   DAC

Almost all circuits used to produce an analog voltage from a multi-digit place-value word are based on the summation of currents. The basic circuit can be adapted to any number base, and has been used extensively with BCD numbers. It is most efficient, however, when pure binary numbers are used. Consider the resistor network in Fig. 7.6(*a*).

Since the two resistors on the right of Fig. 7.6(*a*) have the same resistance, the same current, i, flows in both. These two resistors and the adjacent resistor R can be considered as a single 2R resistor. Thus it can be deduced that the current in the next 2R resistor is 2i, and so on. The current in each leg of the circuit is twice that flowing in the one to its right.

If switches are added to this network to divert these currents to two separate earth points, then the current flowing in one leg will be proportional to the binary representation of the switching pattern and the current in the other leg will be its complement (Fig. 7.6(*b*)).

The current i_0 of Fig. 7.6(*b*) can be precisely converted to a voltage by the use of a *differential input voltage amplifier*, commonly called an operational amplifier. These amplifiers have a transfer function

$$V_0 = A \times (V_+ - V_-)$$

The gain A is typically of the order of 10^5, so that for the output to be within the dynamic range of the amplifier (typically ± 10 V), the voltage difference between the two inputs must be of the order of 0.1 mV. The feedback loop shown provides negative feedback and thus, the V_- input will be clamped at the same potential as the V_+, that is, 0 V (a *virtual earth*, as required for the R–2R ladder). The current i_0 must flow through R_2. The output voltage is simply

$$V_0 = -i_0 \times R_2$$

and is thus proportional to the binary input word and to the reference current input to the R–2R network. The equivalent resistance of the network is simply R, hence, if the feedback resistor R_2 is made equal to this, the output voltage from the circuit of Fig. 7.6(*b*) will be

$$V_0 = -D \times V_{REF}/2^n$$

where D is the binary weighted integer value of the digital input, and n is the number of stages (i.e. bits).

Figure 7.6 (*a*) The R–2R current ladder and (*b*) the binary DAC circuit.

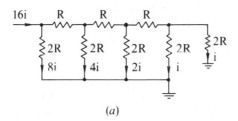

(*a*)

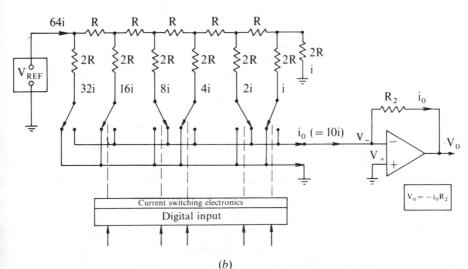

(*b*)

The R–2R network is usually made by microcircuit techniques with thin film, metal resistors on a silicon or ceramic substrate. These are very stable and, since the entire network is made on a single chip, it can be manufactured very accurately and matched, if required, by automatic laser trimming of the individual resistors. If R_2, the amplifier feedback resistor is included in the network, only the resistance ratios and not the absolute values need to be tightly controlled. Since they are in close thermal contact, the resistances will track together with temperature changes and maintain their resistance ratios over a wide temperature range.

Commercial chips are available for 8-, 10-, 12- and 16-bit arrays. (A 16-bit network requires that the most significant resistors are matched to about 20 ppm.) Box 7.3 describes some single-chip DACs that are designed for easy interfacing to a microcomputer.

Box 7.3
DACs for microcomputer interfaces

This box describes a few of the many commercial DAC chips and modules designed for microprocessor interfacing. The reader is referred to manufacturers' data manuals for further information. The most advanced devices are made by the specialised manufacturers, such as Analog Devices and Datel. A wide range of impressive, expensive 'state-of-the-art' chips and modules are available. Manufacturers such as Motorola, Fairchild, National, etc., also offer a good range of devices, usually less advanced, but much cheaper.

8-bit DACs

An 8-bit DAC is easy to interface to an 8-bit databus. Commercial devices range from simple, cheap chips that need considerable logic support, to complete devices that connect directly to the computer busses.

The AD7523 (Analog Devices) chip, and the DAC-08 'industry-standard', are two examples of the simple devices. The AD7523 is a CMOS device with thin film resistors on the chip. The CMOS technology provides current switching from the R–2R ladder via simple ohmic (zero-offset voltage) CMOS switches and allows positive or negative reference voltages to be applied. This class of device can be used for bipolar outputs and as a digital attenuator for a.c. signals.

The DAC-08 chip uses bipolar transistor technology which cannot provide simple current switching, 'current-steering' transistor switches must be used. This restricts the chip to monopolar switching, but external circuitry can be added to provide bipolar outputs. There is an incidental advantage in that the output current has a high compliance; the output voltage can vary without affecting the current. Thus, if a high-impedance output voltage is acceptable, it can be achieved by simply connecting the current output to ground through a suitable resistor. Fig. 7.7 shows typical interface circuits for both devices.

More elaborate 8-bit DAC chips include the databus latch and a voltage reference source. The Motorola MC6890 enhancement of the DAC-08 includes a master/slave 8-bit data latch with a RESET input to clear it, and a precision 2.5 V reference. Extra feedback resistors are included in its network to allow mono- and bipolar output voltage ranges. The chip requires only some address decoding (and an operational amplifier for low-impedance output) to provide a complete DAC function.

The Analog Devices AD7524 chip includes the data latch with two ENABLE inputs (to allow decoding and latch timing), the more elaborate AD558 is a complete 8-bit DAC (including latch, reference and output amplifier), which can operate from a single $+5-+12$ V supply.

Fig. 7.8 shows the simple circuits needed for interfacing to an Apple microcomputer. (The digital data is read while DEVSEL is low; the direct connection with very short propagation delay ensures that the data is latched before the databus changes after the end of the transaction.) New improved devices are continually being introduced. Many of these provide very simple solutions to 'last year's' impossible problem. It is essential to

Figure 7.7   Interfacing the AD7523 and DAC-08 DACs. The LATCH signal could be derived from the logic described in Box 7.1.

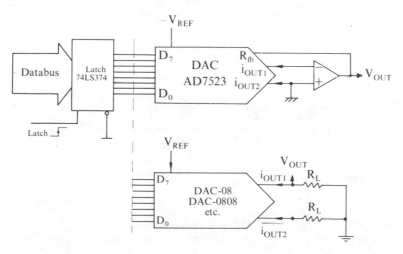

obtain up-to-date data manuals. They also provide all the information needed to interface them into elaborate interfaces or into any microcomputer.

12-bit DACs

For a 12-bit DAC in an 8-bit machine, data must be passed in two words. A minor problem arises when the input data requires that both words be changed; a *glitch* would normally result in the analog output between the two data transfers. This is eliminated by adding double-buffering to the data latches; each word is latched individually into an intermediate latch whose contents can then be latched as a single big word into the DAC. This can be done with external 8-bit data latches and a strobed 12-bit latch, but very nice chips with all the necessary logic and latches are available.

The Analog Devices AD567 and AD667 chips and the National DAC1230 are good examples. The AD567 has an internal precision reference. It has three independent 4-bit latches that allow left- or right-justified 12-bit data entry from an 8-bit databus. A fourth latch transfers the full 12-bit word into the DAC.

The DAC 1230 has a simple but subtle latch structure that only allows left-justified input data, but it is pin compatible with the 8-bit DAC0830 chip. It has no internal reference, but is very cheap. Both chips are extremely easy to interface to any 8-bit microcomputer.

Figure 7.8   Interfacing the AD7524 and the AD558 8-bit DAC.

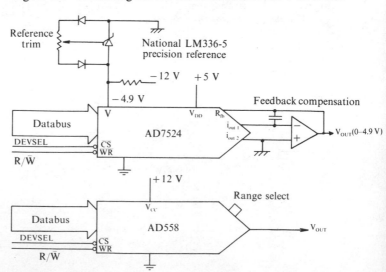

Manufacturer's data manuals provide detailed application information and describe the wide range of devices available. These include ultra-fast 'deglitched' convertors for CRO drive, very high precision and electrically isolated current-mode convertors for delicate instrumentation etc.

7.4.2   ADC

ADC requires some circuitry to quantise the analog input signal (usually a voltage) to its best digital approximation. Several common techniques are used; all are available as commercial devices in modular circuit blocks or as single-chip microcircuits. Each of these techniques is briefly described and their fields of application are indicated in the following subsections.

7.4.2.1 *Successive approximation convertors*

The successive approximation ADC uses a sequential logic block and a DAC such as the R–2R network (Fig. 7.9). The input signal is quantised to its best approximation from the DAC by a sequence of control signals; during this *conversion time* the analog input must be steady. A rapidly changing input signal is usually conditioned by a *sample/hold amplifier* (see Box 7.4). When commanded to start a conversion, the sequence control logic first sets only the m.s.b. of the DAC **true** and, after waiting a short time for settling, it consults the comparator output to see if the unknown voltage is greater or less than its DAC output. That digital bit is either left high or reset on the result of the conversion. It then moves down through the lesser bits to *successively approximate* the DAC output

Figure 7.9   The components of a successive approximation ADC.

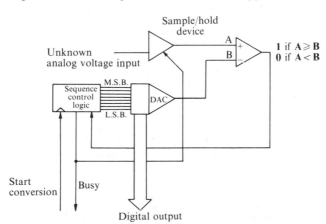

to the analog input. This process is depicted for a 4-bit conversion in Fig. 7.10. The conversion time for this class of ADC rises logarithmically with resolution, hence it is favoured for high-resolution, fast applications. The precision and linearity of the internal DAC determines the performance of the ADC; the DAC invariably uses the well-evolved R–2R network. Nonetheless, characteristic errors of the network impose restraints on some applications. This is explained in Fig. 7.11.

Due to the small but finite errors in the R–2R network, the digital bits will not convert perfectly to their analog equivalent. This is shown on the left-hand (analog voltage) axis of Fig. 7.11(*a*), where the error in each of the DAC bits is shown. These errors add algebraically for every conversion value and result in a response which can deviate considerably from the ideal finite resolution line which is a regular 'staircase' along the ideal response line. The error is kept as small as possible, but, even with just 4 bits as depicted in Fig. 7.11, and with a bit error of only $\pm\frac{1}{3}$ of the least bit, the algebraic accumulation of errors causes large fluctuations in the net conversion error. In the simple case shown, it is possible for the error in two adjacent steps to be such that the response line is not *monotonic*, the next

Figure 7.10   Approximation sequence for a 4-bit convertor.

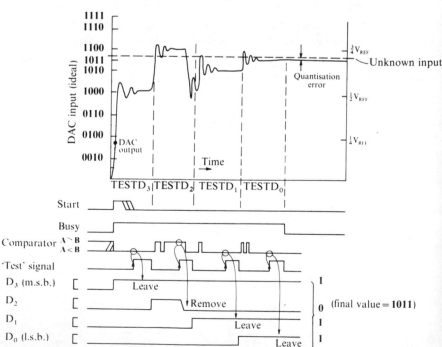

higher logic step can have a lower analog conversion value. The ADC control sequence will then never return a value of this *missing code*.

If a perfectly random analog input signal is converted, the probability of obtaining any code is proportional to the span of that code (the right-hand axis of Fig. 7.11(*a*)). The probability is zero for a missing code.

The probability distribution for the simple example of Fig. 7.11(*a*) is

Figure 7.11   (*a*) Conversion error with the successive approximation ADC, (*b*) probability of detecting codes.

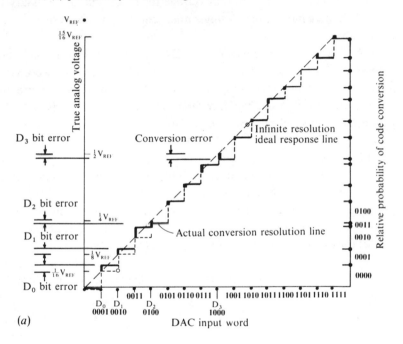

(*a*)

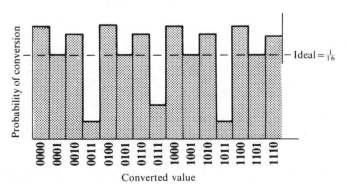

(*b*)

shown in Fig. 7.11(*b*) as a histogram. The ideal expected probability is precisely $\frac{1}{16}$. The effect of small bit errors is very obvious. This is the *differential error* of the convertor.

A technique called *dithering* is often used to reduce the effect. A small random analog offset voltage is added to the input signal, and its digital converted value is subtracted from the conversion value. This averages the error over several adjacent conversion values and can considerably improve the differential error.

Common devices can provide 12-bit conversion in a couple of microseconds, with an internal voltage reference and all necessary timing and bus drive hardware in a single microchip; 16-bit convertors are readily available.

The successive approximation ADC has a very favourable (speed) × (precision) × (cost) product and is recommended where direct conversion is needed, especially in the range of 8-bit to 12-bit conversion. Care must be taken whenever its significant differential conversion error is important, as in a multi-channel analyser (see Chapter 9).

7.4.2.2 *The integrating ADC*

The integrating convertor digitises an unknown voltage by timing how long it takes a constant current to charge a capacitor to the unknown voltage. It requires some simple logic circuitry to respond to a start command, to count a clock signal while the capacitor is charging and to reset the capacitor and the timing counter. Fig. 7.12 shows the elements of a

Figure 7.12 A simplified schematic diagram of a single-slope A/D convertor.

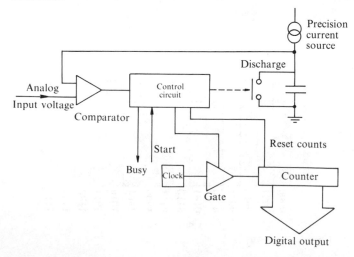

Digital output

single-slope convertor. This simple circuit allows the constant current source to charge the capacitor from zero volts until the unknown analog input is reached; the time taken is proportional to the analog input voltage. Very simple and cheap converters based on this principle are provided within the Apple and BBC microcomputers. (See their reference manuals.) The conversion time is proportional to the analog input voltage. A realistic upper clock rate for simple CMOS devices is about 1 MHz; for 10-bit precision, the conversion time will be about 1 ms.

The more elaborate, *dual-slope convertor* first charges the capacitor at a rate proportional to the analog input voltage for a fixed time (thus integrating the input during this time), and then discharges it at a rate proportional to a fixed reference voltage. The time required (number of clock pulses) for this discharge yields the ratio between the reference voltage and the *integral* of the unknown analog voltage during the charge period. This convertor is, of course, even slower than the single-slope convertor.

Further elaboration to a *quad-slope convertor* allows ground and offset errors to be eliminated and 16-bit precision to be easily and economically, albeit slowly, achieved. The current source can be made stable to a few ppm over the typical conversion time. This provides extremely good differential linearity and avoids the shortcomings of the successive approximation technique.

As mentioned above, the conversion time is dependent on the analog input value. When implemented in sophisticated discrete circuitry and with ECL logic and counters, the clock rate can be raised to about 100 MHz and the precision can be extended to 20 bits (± 1 ppm). At this level of precision, conversion requires about 30 ms at full scale.

These converters are used for slow but very precise conversions or where differential linearity is essential. They are not as well developed for microcomputer applications as are the successive approximation devices (but see Analog Devices AD7550). Their major use is in digital panel meters or in industrial process control; in these applications, their slow speed is no problem and their simplicity is an advantage. They have always been used in precision multi-channel analysers; here their slow conversion speed is a serious disadvantage, but is overridden by the requirement for very good differential linearity.

7.4.2.3   *Flash convertors*

The flash convertor was originally developed for high-speed conversion of radar and video signals. Resolution is quite low (usually 6–8 bits), but their extremely fast conversion time (as little as 10 ns) allows for real-time digitisation and subsequent processing of a video signal. They consist of arrays of resistors and comparators connected as shown in Fig. 7.13. The comparator outputs drive a combinational logic decode array which synchronously generates on command, a binary coded word corresponding to the instantaneous signal level. Linearity is dependent on the precision of the large resistor array and is quite good, the differential linearity is also good. They need no sample/hold since they perform synchronous *comparison*, rather than conversion, of the input analog signal.

Typical single-chip devices provide 6-bit resolution expandable to 8 bits, and 50–100 MHz sampling rates (see Analog Devices AD5010/6020). At this rate they must be driven and processed by ECL circuitry. The slower RCA CA3308 and Thomson EF8308 devices convert at up to 10 MHz, to 8-bit precision and at very moderate cost. These convertors are only available with linear response, but there is no reason why they could not also be provided with logarithmic response.

Figure 7.13   The elements of a flash ADC.

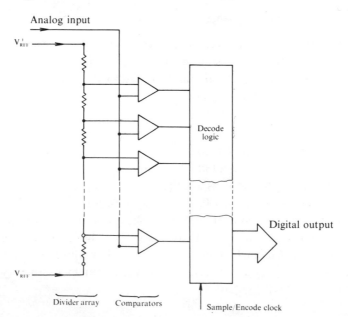

7.4.2.4   *Voltage/frequency convertors*

A *Voltage/Frequency* or *Current/Frequency Convertor* (*VFC* or *IFC*) is simply a voltage or current controlled oscillator. It outputs a binary square wave (clock pulses) at a frequency proportional to the applied voltage or current. The principal use for these devices is to transmit accurate analog signal information over long communication lines. Commercially available circuits allow V/F and F/V conversion so that the original analog signal can be restored at the receiving end of the line.

Chapter 1 explained the importance of binary logic in overcoming signal degradation and noise problems. Problems of noise and stability will always be present wherever analog signals are handled. In the sections above, we have considered ADC systems with typically 12-bit binary resolution, that is, 1 part in 4096. For a 10 V (full scale) signal, the conversion resolution is 2.5 mV. Noise or instability in the system must be kept below this level. In a computer environment, the fast logic signals can induce quite large switching transient signals. Long conductor lines are particularly susceptible both to this noise and to general interference from power mains and from electrical machinery. It is quite difficult to transmit clear, quiet analog signals over any kind of distance.

The VFCs and IFCs allow delicate noise-sensitive analog signals to be converted, at their source, to robust binary logic signals. The analog information is transformed into a robust digital signal in the time (frequency) domain. There is, of course, a trade-off; several cycles of the logic signal are needed to provide any information. The response time of the convertor is thus a few cycles of the clock.

These convertors can be used as ADCs by passing their clock output into either a period timer, or to a frequency counter. Both of these functions are easily set up and are particularly easy to interface to a microcomputer bus. A digital counter which times the period of the VFC provides a direct digital output at the end of every cycle of the V/F oscillator, proportional to the reciprocal of the original analog input. A frequency counter, on the other hand, integrates the original analog input throughout the counting period and returns the average value of the analog input during that period. Both arrangements are shown in Fig. 7.14.

Commercial VFC and IFC chips are available. They are not as precise or as stable as the other ADC devices described above, but they are very effective as an ADC in those applications where their noise immunity or their special conversion characteristics can be used to advantage.

Figure 7.14   Two possible configurations to extract information from a VFC.

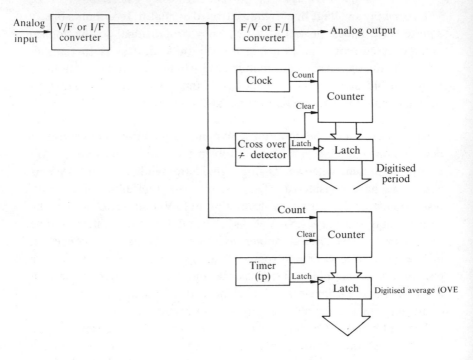

Box 7.4
ADCs and their interfacing

This box gives brief descriptions of a few examples from the wide range of commercial ADC chips and modules, and will explain how they can be interfaced and operated in a microcomputer. The interface hardware is generally more elaborate than that for a DAC, due to the need to initiate and monitor the conversion sequence.

8-bit ADC interface
Fig. 7.15 shows interface logic for a simple, 8-bit ADC using an Analog Devices AD582 sample/hold amplifier and an AD673 ADC chip. The AD673 has an internal clock, its own reference voltage and Tri-State outputs. It has a low-impedance analog input and hence must be driven from a low-impedance buffer; the sample/hold amplifier is suitable.

The data latch and enable signals are derived from a circuit such as that in Box 7.1. The sample/hold amplifier is shown configured as a unity gain follower.

The ADC sequence is started by the latch signal shown (L_x), the positive edge initiates the AD673 conversion. The $\overline{DR}$ output from the AD673 is used to control the sample/hold amplifier. It can also be sent to another input port of the interface (used as a status register) so that the processor can determine when the conversion is complete. In the case of the AD673, the conversion only takes about 20 μs and could be handled instead by a short delay routine. When the conversion is complete, the data can be read from the location defined by the $\overline{OE_x}$ signal.

If several analog inputs are required in an interface, they are usually provided from a single convertor with an analog multiplexor (or switch) ahead of the sample/hold amplifier. Commercial integrated circuits are available for this function; refer to the Analog Devices AD7501–2 multiplexors and to the National ADC0808/0816 multiplexed ADC or their equivalents. The National ADC0844 is worthy of special mention; it is a very versatile and efficient multiplexed 8-bit ADC. The AD7574, AD7581 and their equivalents from other manufacturers should also be examined.

10- and 12-bit ADC interfaces

These convertors are similar in application to the 8-bit devices. The principal difference is that the converted data must be read in multiple words. For a 12-bit convertor with direct logic inputs, it is usually

Figure 7.15   A simple ADC interface. Note the self-latching enable signal of the 74LS373 status register. $\overline{DR}$ (Data Ready) is analogous to the 'Busy' signal in Figs. 7.9 and 7.10.

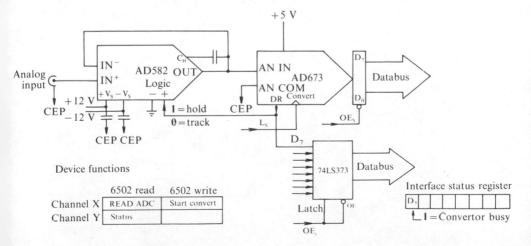

convenient to read the top eight bits through a dedicated input port and to read the extra four low-order bits through a further input port, or through a status register, as used in Fig. 7.15. Fig. 7.16 shows a typical arrangement.

Microcomputer compatible 10- and 12-bit ADC chips have special provision for passing the conversion data onto the databus in two independently enabled words. This yields simple interface control logic circuits. The Analog Devices AD574A is a good example of a very convenient chip.

Other devices

The range of devices available is just too great for even a brief mention here. The interfacing of integrating convertors, flash convertors and VFCs has not even been discussed. The information in this box and in Box 7.1 should, however, aid a soft entry into the interfacing of analog conversion devices.

Caveats

Before attempting to design or build any hardware, it is important to realise that the inside of a computer is a very noisy place. Analog lines should be kept short and well shielded by ground planes; all analog chips

Figure 7.16   A suitable circuit to read data from a 12-bit ADC over an 8-bit databus. Note that only the output logic is shown.

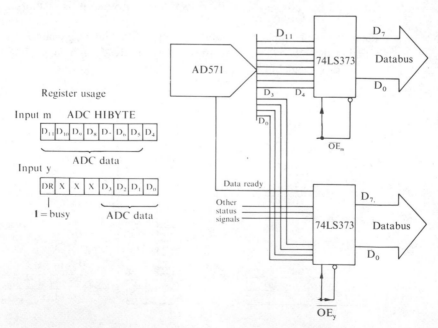

should be decoupled to their power supply ground line very close to the chips. A rigorous *Central Earth Point (CEP)* philosophy is essential; special care must be taken to avoid digital signal ground return noise.

7.4.3 Analog signal manipulation

The DAC and ADC chips discussed briefly in the previous section can be configured and interconnected to provide versatile signal manipulation systems. Many of the DAC chips can be used as multipliers, or more correctly as digital attenuators. If a variable analog signal is used as the reference voltage the output will be attenuated by the digital input data (Fig. 7.17(a)). Some DAC chips allow four quadrant multiplication, with signed inputs. Care must be taken with analog feedback compensation in the DAC; consult the manufacturers' data manuals for more information.

The block circuit shown in Fig. 7.17(b) allows the conversion gain of the ADC to be controlled by the digital data presented to the DAC. The reference source of the AD574A is used as the reference input for the AD567, but the output of the AD567 is used as the reference input for the AD574A. An application of this is described in Chapter 9.

Instrumentation amplifiers and non-linear amplifiers allow enhanced conditioning of analog input signals. A wide range of instrumentation amplifiers are available. They allow simple gain change, signal bandwidth manipulation and high gain with high stability. ADC chips are available with inbuilt conditioning for low-level signals (refer to the Analog Devices AD670).

Figure 7.17   Analog signal processing with DAC and ADC circuit.

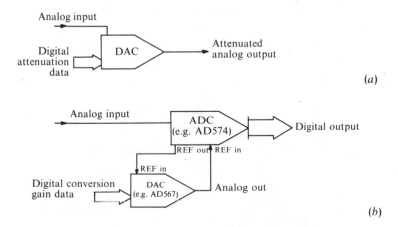

A wide dynamic response from an ADC normally requires a large word-size to provide adequate resolution, but fine absolute resolution is only needed at the low end. The Analog Devices AD7110/AD7111 chips, or their equivalents, provide a *logarithmic* response to their digital (attenuation) input. These are intended for audio attenuation, but can be used to expand the dynamic conversion range of an ADC.

A wide dynamic range can also be achieved with a moderate resolution ADC, by conditioning the input signal through a logarithmic preamplifier. These are easy to build from operational amplifiers, but there are, of course, very good commercial devices available. The Analog Devices 755 module provides up to six decades of input dynamic range.

7.5   Data collection from counters and timers

The problems of ADC can be avoided when outside-world data can be collected directly in the form of binary signals. The inherent advantages of binary logic circuits, high speed, high noise immunity and virtually unlimited precision can then be fully exploited. VFC is a special case of this (Section 7.4.2.4), but there are two other data-collection schemes of special significance, event counting and time interval measurements.

Event counting is straightforward below about 20 MHz, using LSTTL counter chips. ECL devices used for the first few counter stages extend the maximum rate up to 100 MHz without much trouble. Time interval measurement reduces to simple counting when a suitable precise clock is available. The system clock in most computers is crystal controlled and has a stability of about 10 ppm when at a reasonably stable temperature. This provides measurement to 16-bit precision, but it is fairly slow; a 16-bit count takes 65 ms.

The moral is clear; high-precision noise-free measurement is readily available and very easy if time domain or event-counting data collection is suitable. Some simple examples are described in Box 7.5, where simple timing and event-counting circuits are described. Some applications are described in Chapter 9.

Box 7.5
Simple counter/timer interfaces

Commercial VIA chips can provide some of the functions described in this box. We will continue, however, to describe interfaces based on TTL chips, to lead on to the specialised data-collection interfaces described in Chapter 9. PIA and VIA chips will be discussed in Chapter 8. Fig. 7.18 depicts a multi-word counter using the control signals of Box 7.1. It uses 74LS590 8-bit three-state synchronous counter/latch chips (Fig. 7.18(a)). These have an 8-bit binary synchronous counter with fast carry propagation and an integrated 8-bit latch and three-state bus driver. Fig. 7.18(b) shows a cascaded array of these counters arranged as a digital timer, driven by the computer system clock. The clock timing is such that the counter propagation delays occur during the address phase of processor

Figure 7.18   (a) The 74LS590 chip; (b) several 74LS590 chips assembled into a multi-bit counter interface.

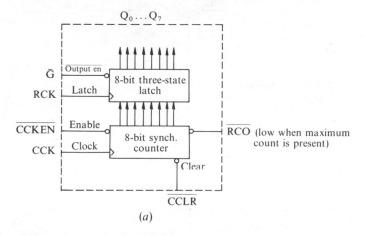

(a)

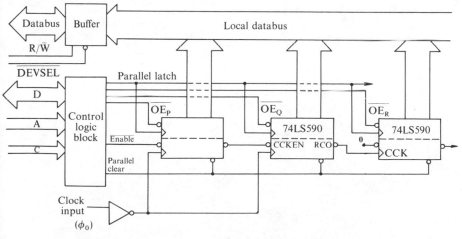

(b)

transaction cycles; it will thus always be stable when latched during the data phase.

The control logic block can be designed to perform whatever timing functions are desired. An example is shown in Fig. 7.19(a) where the logic is arranged such that the counter can be cleared by setting D_0 high at the

Figure 7.19 (a) A multi-word counter interface control scheme, and (b) a simple enhancement.

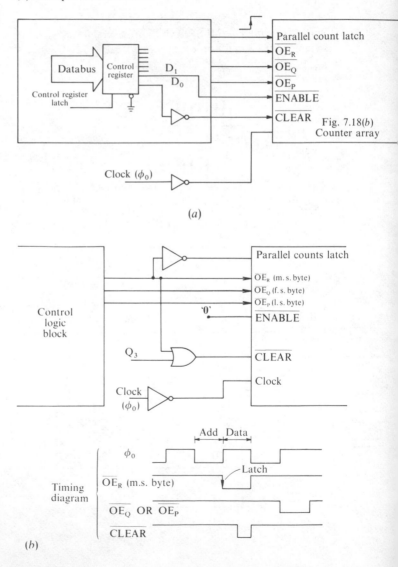

interface control register; the timer can be enabled by setting both D_0 and D_1 low.

At any time, the entire counter contents can be latched simultaneously by a processor write to an address used to strobe LATCH COUNTS, but the three latched data words (most significant (m.s.), fairly significant (f.s) and least significant (l.s.)) can be read later at the program's leisure.

Fig. 7.19(*b*) shows an enhancement, with the LATCH COUNTS strobe generated just before the most significant word is read, and the common CLEAR pulse is generated just after the latch strobe. This creates a simple interval timer which latches the number of clock pulses since the prior latch (like a lap timer).

The counter control circuitry can be further enhanced to select either clock pulses or external events to be counted. External signals can be used to start and stop the counter. Design of suitable circuitry for both of these options is left as an exercise.

When the counter is controlled from external devices, the program must have some means of controlling it and of knowing when the counter is running or when it has stopped with valid data present. This can be provided easily by flags in the interface control and status registers. When counting external signals, care must be taken to avoid timing conflicts with the data latch signals. The control logic must freeze the counter before latching, to allow propagation delays to settle (about 50 ns for the 74LS590). Design of this circuit is left as an exercise.

The circuits described above can carry out all of the data collection needed for the VFC period and frequency measurements described in Section 7.4.2.4.

Further sophistication can be provided with the 74LS593 counter chip. This chip is like the 74LS590 but is enhanced with an 8-bit latched-input register which can be used to preset the counter. The input register can be latched directly from the databus and then transferred into the counter. The counter will overflow after the time interval (or event count) determined by this. A simple connection from the m.s. byte overflow output to the counter preset, reloads the preset data automatically and yields a programmable divider on the counts input. This can be used for interrupt generation etc.

Multi-bit UP/DOWN counters with three-state bus buffers are also available. These can be used in many applications involving bi-directional motion, such as machine controller, tracker-ball or mouse interfaces. (See Chapter 9.)

7.6   Interrupt interaction

Section 7.3 described the interaction scheme which required the program to interrogate or poll the interface status ports from time to time. *Interrupt interaction* provides a more efficient interaction scheme. The outside world can interact directly with the processor over direct lines in the *control bus*.

It is rather like the emergency communication cord of a train. The direct line in the computer's control bus allows any device to signal to the processor, without program intervention, that it needs attention. The line is shared by all devices, hence can give no immediate information about the caller; the processor's response must determine this.

This section will describe the extra computer and interface hardware needed for this type of interaction, and will then describe the techniques used in *interrupt service routines* to determine which device or devices have requested service.

When the processor reaches the end of an instruction execution cycle, and is about to fetch the next instruction opcode, the contents of its registers completely define its state. If the register contents are saved somewhere in memory, program execution can be diverted to another routine. When the register contents are restored to their previous state, the processor will resume execution of the interrupted program from its prior state. It is possible therefore to interrupt a program at the end of any instruction execution cycle, and to divert the processor to an interrupt service routine. At the end of this service routine, the interrupted program can be resumed unaffected.

In this section, we will explore how this is implemented in a very simple manner in the 6502 processor, and will then outline some powerful and sophisticated implementations in other microprocessors and computers. We will start by describing a closely related feature of computer operation which provides an automatic bootstrap start-up (Sections 6.6 and 8.4.3). After discussing the hardware interrupt, its extension to the notion of *software interrupts* will be explained.

7.6.1   RESET response

Section 6.6 showed that a microprocessor must always be executing valid program code. This raises the problem of how it ever manages to start. Some provision obviously must be made to force the processor to start from some fully defined state; this RESET provision will now be explained.

The RESET input to the 6502 processor is a simple logic signal line which is normally held high. The computer hardware must, at power-up or whenever required, pull the input low long enough for the power supply to come up and for the clock to stabilise. The RESET line generates the simplest interrupt response of the 6502.

A processor's response to any interrupt control signal is similar to its response to an opcode in the IR; the processor executes the instruction interpretation sequence of control signals (Chapter 5). Interrupt response can be considered as a *pseudo-instruction*.

When RESET is pulled low, the 6502 stalls; when it is returned to logic high, the processor begins a control signal sequence equivalent to

JMP ($FFFC)

It loads PCLO and PCHI with the data at $FFFC and $FFFD respectively, and fetches the next opcode from the address thus defined.

This pseudo-instruction is the one that allows the computer to boot-up 'from cold' in an orderly manner; the first fetched opcode will come from a fully defined address. Obviously, the indirect addresses and the program code pointed by them must be in permanent memory (ROM) so that they are available at power-up. Note that the RESET interrupt line causes an address defined *internally within the 6502 logic* to be accessed but uses this as an indirect address pointed into the system software. Thus, the response of the processor is hardware defined but can be software controlled. This general concept is called *interrupt vectoring* (Fig. 7.20); we will return to it later in this section. In the Apple computer, the power-up RESET vector is in the Monitor ROM, in the BBC machine, it is in the MOS (Machine Operating System) ROM.

7.6.2   External hardware interrupts

The 6502 has a very simple interrupt structure, with only a *Non-Maskable Interrupt* (*NMI*) line and a single *maskable Interrupt ReQuest* (*IRQ*) line. The NMI will be described first.

Figure 7.20   The concept of a hardware vector.

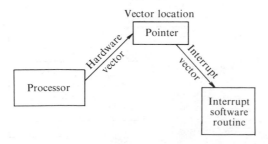

7.6.2.1 *NMI*

The NMI interrupt sequence is initiated by a negative-going edge on the NMI control input (Fig. 7.21). This edge triggers a flip-flop inside the 6502; at the end of every instruction cycle, the 6502 tests (and clears) this flip-flop. If an NMI was sensed, the 6502 invokes the NMI pseudo-instruction control sequence. The effect is as follows.

(1) The PCHI contents are pushed onto the STACK.
(2) The PCLO contents are pushed onto the STACK.
(3) The P Register contents are pushed onto the STACK.
(4) The I bit of the P Register is set high.
(5) The control sequence for

JMP ($FFFA)

is executed.

This stores the processor status, and the address of the next opcode in the interrupted program, on the STACK and then diverts the processor to the routine vectors through $FFFA, FFFB. This sequence does not save the contents of all the registers, it only saves those which are changed by its execution. (The reason for the change in the P Register will become obvious below.) The programmer must ensure that all registers to be used in the interrupt service routine are moved to temporary storage and

Figure 7.21   Hardware, processor and service interaction for NMI service.

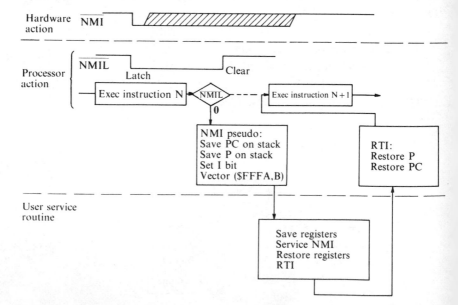

recovered at the end of the routine. Use of the STACK provides reentrant interrupt code (refer to Box 6.3), but stack access is rather slow and space is limited.

The interrupt routine is ended by the RTI opcode. This is similar to the RTS described in Chapter 6, except that the P Register is fetched from the STACK before the return opcode address. Provided that all processor registers have been restored and that the STACK has not been abused, the processor will resume execution of the interrupted program.

7.6.2.2 *IRQ*

The processor response to the IRQ interrupt line is very similar to that from the NMI line, but the IRQ line is level-sensitive (not edge-sensitive) and the processor will not respond if the I bit of the P Register is high (Fig. 7.22). This second feature allows the processor response to be *masked out* (disabled) during an interrupt service routine or by the software instruction (SEI), and to be enabled by the CLI instruction. (Cf. the *non-maskable* NMI.)

The 6502 response to the IRQ level is as follows.
    (1) PCHI pushed onto the STACK.

Figure 7.22   Hardware, processor and service interaction for IRQ service.

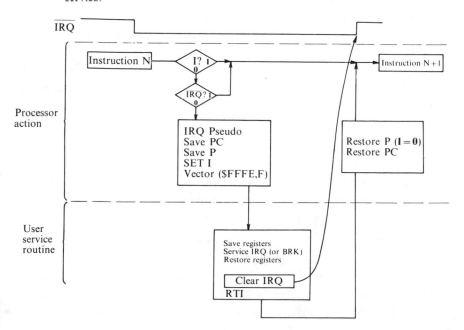

(2) PCLO pushed onto the STACK.

(3) P Register pushed onto the STACK.

(4) I bit of P Register set high.

(5) JMP ($FFFE) is executed.

When the IRQ pseudo-instruction is executed, the I bit is set high to prevent further IRQ response. This allows the control line to have a *level-sensitive* response and consequently allows multiple interface devices to pull the line low concurrently for service requests. Handling of such multiple request situations is described in Section 7.7.

At the end of the service routine, the RTI instruction described above will restore the I bit to its low state and will thus automatically enable further interrupts. It is essential, of course, that the service routine forces the calling device to remove its IRQ signal before the RTI is executed. If several devices are requesting simultaneously, the persisting IRQ signal will force an immediate reentry to the interrupt routine (more about this later).

7.6.3   Software interrupts (BRK)

The concept of a vectored interrupt allows an external device to directly access service routines via a software defined vector table. This same concept can also be used within the machine to allow simple access via a simple vector table, from a high-level program to low-level service routines for the computer's peripheral systems (VDU, disc, etc.). These routines are usually part of the computer's *operating system*. The interrupt return allows the high-level program to resume after the service routine is finished.

Another application of this scheme allows difficulties encountered within a user's program (such as an arithmetic over/underflow), to be handled by invoking an error handling routine accessed, like an interrupt, from within the high-level language. This is commonly called an *error trap*.

These expanded features can be implemented by expanding the interrupt concept to include *software interrupts*, invoked by program instructions. The new 16-bit microprocessors invariably provide very powerful software interrupt facilities; our humble demonstration machine, the 6502, has only one example, the *BReaK execution* instruction (BRK).

The BRK instruction of the 6502 (OPCODE: $00), invokes a non-maskable interrupt sequence similar to the IRQ response and via the same vector ($FFFE, FFFF) (Fig. 7.23). The B flag of the P Register is set high *before* its contents are put onto the STACK. The Apple firmware handles this interrupt by displaying the contents of all the processor registers on the screen. (This is very convenient for debugging programs, since the BRK command can be placed at strategic points in a routine, and the processor state is returned when and if it reaches that instruction. It is a simple and convincing demonstration of the power of software interrupts.)

Box 7.6
Apple II interrupt firmware

This box describes the Apple firmware and hardware for servicing the four 6502 interrupts. This is to demonstrate the basic requirements of interrupt service software. More elaborate interrupt servicing with the 6502 and other, more powerful processors, will be discussed in Section 7.7.

The interrupt vectors of the 6502 memory map are at the top of address space. This area must be occupied by ROM, to provide valid code for boot-up via the RESET vector. The top 2K addresses are used in the Apple, by

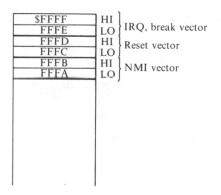

the system MONITOR and the service routines for all inbuilt I/O functions. A full description of this is provided in the references at the end of this chapter.

The vectors in ROM cannot normally be modified by user software, hence the NMI vector must be revectored from there into RAM, where the user can enter the interrupt service code.

The shared use of one vector by the IRQ and BRK interrupts requires that the interrupt service routine has an overture to distinguish the two interrupts; the overture is embedded in the MONITOR ROM.

The Apple II Plus interrupt service routines are outlined below.

Vectors
NMI: $FFFB,FFFA:- $03FB
RES: $FFFD,FFFC:- $FA62
IRQ: $FFFF,FFFE:- $FA40

Service routines

NMI

The vector points the processor at the location $03FB at the end of the 'free' PAGE #3. There is just enough space there before the start of the TEXT screen buffer (PAGE #4– #7) for a three-word instruction, which revectors the interrupt to the user's routine:

JMP $aaaa

or

JMP ($aaaa)

RESET

The vector points the processor into the initialisation routines in ROM.

IRQ and BRK

The vector points the processor into the handling overture starting at $FA40 (see also Fig. 7.24).

Figure 7.23   BRK processor response.

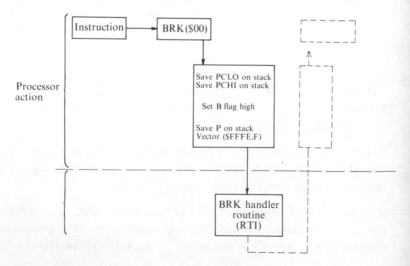

```
FA40: 85 45    IRQIN  STA $45      :Save accumulator contents
      68              PLA          :Get P off stack into A Register
      48              PHA          :Push back to preserve Stack
      0A              ASL A        :Three left-shifts in A
      0A              ASL A        :to get B flag
      0A              ASL A        :into m.s.b.
FA47: 30 03           BMI BREAK    :if B then BRK
      6C FE 03        JMP (USRIRQ) :if not redivert to user vector
FA52  28        BREAK  PLP         :Here begins the MONITOR
                                    routine
```

The overture vectors the IRQ response via the address defined by $03FE, 03FF. The user interrupt service software must load that address pair with the start address of the actual interrupt service routine.

Note that the overture needs to use the A Register for testing. It saves the old A Register contents at a reserved address; any user interrupt routine must rescue the old contents from that address; if reentrant interrupts are required, the old A must be recovered and put on the STACK.

The Apple MONITOR uses the BRK command to invoke a register-contents display routine and then to drop into MONITOR at command level.

Figure 7.24 Apple IRQ service overture.

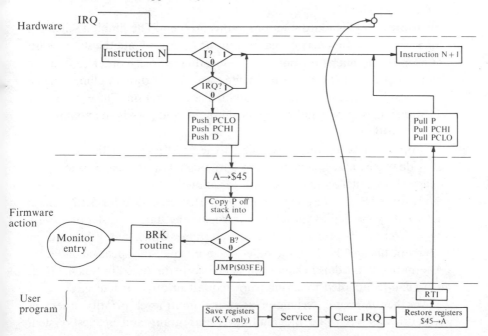

The IRQ pseudo-instruction requires 8 machine cycles, and the simple overture imposes another 26 cycles overhead; the IRQ response thus requires about 34 μs before the processor is in the interrupt service routine. The NMI response, being direct to the user's code, requires 8 cycles to respond and three more to jump to the user routine. Neither of these is very fast!

7.7   Interrupt and priority schemes

This section describes some simple schemes that enhance the primitive interrupt structure of the 6502. The interrupt structure of some other microprocessors are also reviewed. Most microprocessors have a fairly simple 'on-chip' interrupt structure to save space for more crucial processor functions. Support chips are usually available which work in conjunction with the microprocessor to expand the structure to quite powerful multiple priority level interrupts.

The general problem of servicing multiple interrupts is much the same as that described briefly for polled program interaction (Section 7.3). There must be some scheme by which

(1) the individual interrupts can be handled in order of their urgency or priority,

(2) the relevant service routine can be accessed for each possible interrupt.

There are many standard solutions to this problem. The simplest scheme is for a single interrupt service master routine to interrogate the status word of all possible interrupt sources in descending priority order to find the requesting device. This is virtually identical to the status polling scheme discussed in Section 7.3 and needs no further description. The provision of all status flags at one address again simplifies and speeds the polling and priority process.

Support chips for the 6502 such as the 6520 PIA and the 6522 VIA provide this by allowing up to seven status signals at the one port to signal different external conditions and to allow individual programmable interrupt services in response to the single IRQ passed to the 6502. This is achieved by making the program interrogate the status register of all VIA or PIA chips in priority order, to determine which has generated an interrupt (flagged by D_7); the other seven bits flag the specific condition. Alternatively, the daisy-chain priority structure discussed in Box 4.3 (Fig. 4.14) can be adapted to an interrupt request priority structure.

Another common technique used for multi-level priority interrupt structures is very like the multi-level bus grant and request system described in Box 4.3 (Fig. 4.15(b)). A table of vectors can be set up in memory and used to provide the starting address of the service routine corresponding to each level of priority. Support chips to provide this type of interrupt priority structure are available for most 16-bit microprocessors (see, for example, the 8259A chip for the Intel 8086).

The most common, versatile structure is the direct, hardware-implemented vector interrupt. This requires extra logic at the peripheral to provide a pointer to the interrupt service address associated with the interrupting device; it provides fast, direct access to many service routines. The Z80 microprocessor, with its orientation to I/O, has been provided with a fairly elaborate example of this.

7.7.1   Z80 interrupt schemes

The Z80 has four distinct interrupt responses, involving one NMI input line and one maskable interrupt line (INT). The INT response can be programmed to respond in one of three interrupt modes called IM0, IM1 and IM2. The NMI response and the IM1 response (to the INT signal) are very like the 6502 NMI and IRQ respectively.

The IM0 response expects the interrupting device to provide the first interrupt instruction. The processor signals the external interrupt hardware to present this by outputting M1 and IOREQ simultaneously. (M1 signals that the processor is fetching an opcode, the IOREQ commands I/O to supply it.) The Z80 has two suitable instructions in its set, the RST and the CALL. The single-word RST instruction is easily invoked as a single data write from the peripheral and contains a 3-bit pointer; the processor saves the present PC contents on the stack and then uses the RST pointer to load the PC with one of eight assigned zero page addresses; this allows the interrupt to vector to one of eight distinct starting addresses on PAGE #0.

The CALL instruction requires three words and requires, therefore, quite fancy peripheral logic; all three words must be generated on demand from the processor. However, it provides an interrupt-initiated subroutine jump to anywhere in memory.

The IM2 mode is the true vectored interrupt. When the Z80 responds in this mode, it saves the current PC on the STACK and expects the interrupting device to provide a 7-bit data word (the l.s.b. is ignored and assumed as 0). This data is appended to the Interrupt Page Register (the I Register) contents within the Z80; this specifies a two-word (16-bit) address pair in the I page. The Z80 then uses the pointed address pair for an indirect jump. Thus, 128 separate, interrupt vector entry points can be defined anywhere in memory with very quick access. The addition of an external daisy-chain priority structure onto this, yields a very powerful interrupt system at moderate cost.

This scheme of vectored interrupts is available on most other microprocessors.

7.8   DMA (Direct Memory Addressing) interaction

The versatility of the computer is achieved by its *sequential* logic operation. The trade-off for this versatility is execution speed, since sequential operations can never achieve the ultimate speed of the logic elements used. Whereas the memory hardware of the computer can perform a memory operation in one machine cycle, the minimum time in which an 8-bit microcomputer can transfer a single word from an I/O port to a memory location under processor control, is about six bus cycles. Programmed and interrupt interaction both impose considerable time *overheads* to service data transactions between the outside world and the computer. Maximum data transfer speeds can only be achieved therefore, by an I/O structure which operates without processor intervention to allow *DMA*. Such a structure must be able to assume the role of a *bus-master* and to operate both the address bus and the control bus directly. It must also be able to share the busses with the processor in an orderly manner. (See Box 4.3.) It requires considerable extra hardware and can only be justified where the extra speed is essential for computer–world interaction.

The principal conventional use for DMA data transfer is to transfer blocks of data between mass storage systems such as tape or disc and the computer's main memory. This application of DMA allows the transfer to be handled as a background task while the processor is executing instructions.

In mini- and mainframe computers, the processor is usually organised such that it only assumes a bus-master role when it needs to communicate with memory or I/O elements; while it is executing an instruction involving internal registers only, it surrenders the busses. During these free times, another potential master can gain access to the busses for simple data transfers between a peripheral and the computer memory. A DMA peripheral system intended for *block data transfers* typically has control registers which can be loaded with the starting address in memory and the word count. When activated, it will steal bus cycles and carry out transactions without any further processor intervention or bus-time 'overhead'.

Such a transfer scheme can be enhanced by adding a priority structure to allow crucial transfers to be carried out uninterrupted or to allow multiple DMA block transfers to be carried out concurrently.

This powerful and efficient time-sharing scheme is not generally available with simple 8-bit microprocessors since these are usually intimately connected to the computer busses and cannot be disconnected without being stalled. (The more powerful 16-bit microprocessors have some implementation of a bus-sharing DMA architecture.)

The Z80 has BUSREQ and BUSACK control I/O lines which allow external logic to force the processor off the busses (see Chapter 8 about DMA controllers), but the 6502 has no provision at all. In the 6502, therefore, a DMA capability must be provided by logic external to the processor. This is described in Box 7.7.

DMA is commonly used for block data transfer between peripherals and memory, but the technique can be modified to provide very fast interaction for data-collection and control schemes. Fairly simple logic circuitry can be designed to provide very powerful, fast data-collection systems within a simple microcomputer. Some examples are described in Chapter 9.

These enhanced systems still have serious shortcomings which prevent the ultimate power and speed of digital logic and fast memory chips being achieved. One of these shortcomings is the limited memory space of an 8-bit computer, which is too small for very large data arrays. Another shortcoming is the 8-bit data word which is too small for most numerical databases; multi-word-variable transactions require multiple bus cycles and much of the speed advantage of DMA is therefore lost. Both of these problems can be overcome by other techniques to be described in the following sections; some applications are described in Chapter 9.

Box 7.7
DMA provision in 8-bit microcomputers

---

(1) Z80 DMA

The Z80, with its I/O orientation, has been provided with bus control handshake lines. The scheme is implemented as described in Box 4.3, but the internal design of the Z80 prevents interleaving bus access between the Z80 and a DMA peripheral system. DMA access usually needs to stall the Z80. When the BUSREQ is pulled low by an external device, the Z80 responds at the end of the current machine cycle by stalling, floating the address and data busses and responding with BUSACK low; the external device must wait for this response before using the busses. When the BUSREQ is returned high, the Z80 will resume normal operation.

A Z80 instruction cycle is depicted in Fig. 7.25. After the Z80 has addressed memory and fetched the instruction opcode, two clock phases are required for internal logic functions. During this time the Z80 provides a refresh phase for dynamic RAM memory (refer to Sections 2.5 and 8.3). A refresh counter within the Z80 outputs a 7-bit address on A_6–A_0 of the address bus to perform dummy READ cycles through all elements of the dynamic RAM.

A DMA controller chip is available for convenient DMA implementation on the Z80. This chip is block transfer oriented and has internal registers to hold source and destination addresses and block lengths. It can be programmed to operate in four modes:
 (*a*) single-word transfer on request,
 (*b*) burst mode, where words are transferred sequentially whenever the sender/receivers are ready,
 (*c*) continuous mode, where the Z80 is locked out until the data transfer is complete,
 (*d*) 'transparent' mode, where the DMA controller steals the Z80 refresh cycles for DMA transfers.
If the computer system has dynamic RAM and uses the Z80 refresh facility

Figure 7.25   Z80 instruction cycle timing.

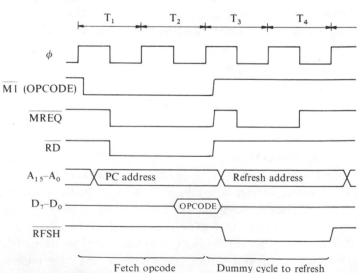

then, when the Z80 is stalled, the external logic must take over the refresh function. (Many short DMA bursts can be used to allow refresh from the Z80.)

The DMA controller chip is oriented towards block data transfers and is not particularly suitable for data-collection interface hardware of the type relevant to this book. However, the Z80 control structure makes external DMA logic quite easy to design and implement; it is only necessary to replicate the control bus timing generated by the Z80 within the external logic for free and direct access to the computer system memory. An application of this is described in Chapter 9.

(2) 6502 DMA

The DMA provision in the 6502 is very simple; there is none! Neither is there any provision for memory refresh. Provision of both of these is left to the computer hardware designer. A simple and easily described example of this is provided by the Apple II hardware architecture.

The 6502 data transaction and instruction execution timing is such that the databus is never needed during the address phase of the transaction cycle. If the computer memory can be fully accessed within the data phase period of 500 ns, then the memory is actually free for other operations during the address phase. This period is used in the Apple for generation of the video display. The normal (internal) Apple display is designed to provide a 7 MHz (approx.) pixel rate. With normal television line scan rates, this allows about 280 pixels across the video screen (and thus supports 40 characters with a seven-pixel character width). At this pixel rate, the Video Display Logic (VDL) requires one 7-bit data word every 6502 clock cycle. The Apple hardware layout allows this by time-multiplexing RAM access via an address bus multiplexor, between the VDL and the 6502.

The VDL accesses the RAM during every address phase to steal a single word from memory. The 1 MHz master clock of the Apple restricts the video display thus generated to a 40 × 24 character screen; quite different, independent circuitry is required to provide higher screen resolution (80-column peripheral cards). The VDL generates the required display address and sequences its address access such that the RAM is automatically refreshed. The block functional diagram of the arrangement is shown in Fig. 7.26.

The 6502 must be forced off the databus during every address phase, so that the RAM can be read by the VDL. This is achieved by a bidirectional buffer between the 6502 and the databus (Fig. 7.27). The buffer is driven from the ϕ_1 clock output so that the buffer reads from the bus during every address phase, and therefore cannot write to it. Provision for DMA access is made by interposing another buffer between the 6502 and the address bus so that it can be isolated or 'floated' for external operation. The control line

Figure 7.26   Apple hardware architecture.

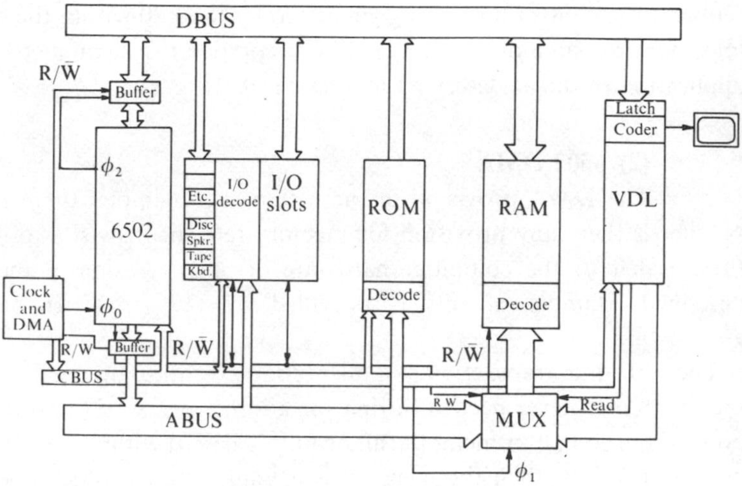

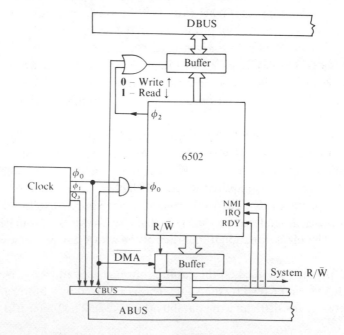

Figure 7.27   DMA provision in the Apple.

used for this buffer also isolates the R/$\overline{\text{W}}$ output of the 6502. The control line (DMA) when pulled low also stalls the ϕ_0 clock input to the 6502. This in turn forces the ϕ_2 output high and floats the databus.

The DMA control line of the Apple stalls the 6502 and disconnects it from the bus system so that external DMA logic can take over. The system clock is left running, however, and VDL access continues to refresh the display and memory during external DMA access. This frees the external logic from providing the memory refresh logic but requires that the external logic act synchronously with the system clock and that it keeps off the databus during the address phase.

There is no handshake protocol; external logic must ensure that the DMA control line is only changed during the address phase (when the 6502 can be safely stalled/disconnected or restarted/reconnected).

There are no DMA support chips for the 6502 but, within the restraints imposed by the description above, quite simple external logic can be designed to allow direct data transactions with the Apple's memory. But all DMA systems stall program execution during DMA access cycles.

Fig. 7.28 shows a block schematic diagram of such a system. The system is controlled by a normal I/O control and status interface, but is able to

Figure 7.28   DMA access logic for the Apple II.

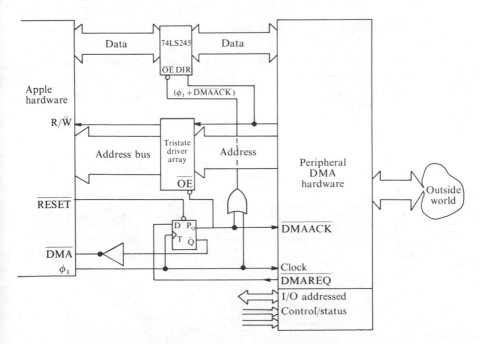

request bus access and act as a *bus-master* when the request is acknowledged (DMAACK). The control logic depicted between the computer and external logic blocks satisfies all of the timing and access restraints described above. Some further examples are described in Chapter 9.

If static memory only is used in a 6502 system, or if the system architecture is designed from the start for DMA access, it is possible to provide fully transparent DMA access; the writer knows of no machines where this is done.

The BBC micro has no external DMA provision although it does use a similar scheme for its VDL (see Section 7.9).

7.9   Video display controllers and linked systems

The VDL is often the most time-critical element in a microcomputer. Resolution is dictated by the maximum data word rate possible, and the operation must be uninterrupted. To avoid heavy bus loading from the VDL, microcomputers generally use specialised CRT or VDU controllers chips. They are available in two general classes. The BBC machine uses the 6845 CRTC, one of the simpler class; it is rather like the Apple VDL, with time-multiplexed databus access between the 6502 and the CRT controller and providing incidental dynamic memory refresh (Fig. 7.29). The BBC, like the Apple, stores video information in normal memory where it can be directly accessed and modified by the program. This scheme limits both the ability of the machine to support DMA, and the display resolution. The 2 MHz 6502A microprocessor of the BBC allows better video resolution than the Apple. The second class of CRT controller is more elaborate and powerful. The processor and the display controllers are

Figure 7.29   The BBC video display scheme.

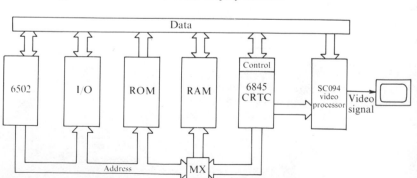

arranged as two linked systems. Display and video control/status information is passed through normal I/O ports on the computer bus. The display controller has a separate bus structure and memory space; this increases the complexity but eliminates all design compromises from the computer bus system.

The elaborate architecture confers considerable advantages. The main computer system needs no reserved memory area for video information, there are no timing and access restraints imposed by multiplexed VDL. The CRT system memory can be organised and timed to suit the video display signal generation. The controller scans this memory (or parts of it) to generate the composite video signal. The display rate is independent of the microprocessor timing.

A typical arrangement is shown in Fig. 7.30. Communication of control and display data from the computer to the CRT controller system is by sequences of words passed by the processor into the FIFO (First In First Out) communication buffer. The controller can accept them during line or frame flyback intervals or, in some controllers, during video data transfers.

The CRT controller has a defined structure for instruction code words and their operands or display data. This instruction/data structure allows the flexibility of instruction execution to be employed by the CRT system; this allows virtually unlimited expansion of CRT controller functions. The NEC 7220 GCD graphics display controller chip uses this scheme to provide a very powerful set of high-level graphic operations. DMA controllers can also be used with this type of system, to allow direct transfer of display data between the computer memory and the video memory.

The interaction scheme of the CRT controller introduces a powerful model for fast systems. A similar but simpler architecture, more easily built from

Figure 7.30   CRT controller scheme with ported communication.

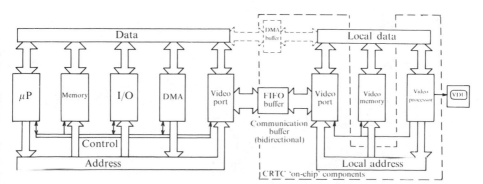

integrated circuits, is described in the next section. This simpler architecture is very well suited to data-collection and control peripherals.

7.10   Dual-ported memory

The architecture of the CRT controller is too complex for implementation in a one-off design. Its development can only be justified for commercial, high-volume systems. Furthermore, the problem of efficient exploitation of high-speed digital logic cannot be solved by replacing the computer bus system with a similar peripheral bus system. The *dual system* concept can, however, be modified to a simple architecture that is well suited to some powerful and fast data-collection schemes.

A scheme called *Dual-Ported Memory (DPM)* is often used in computer architectures with two or more main processors. A block of memory is arranged such that both processors can access it; the concept of shared bus access between two masters is inverted here such that the two (or more) processors can run independently within their own memory spaces, but the shared memory array can be addressed by the two separate bus systems (Fig. 7.31).

Arbitration/priority logic is needed for orderly shared access to the common memory; VLSI chips are available! (See Section 8.2.2.) The scheme can be extended such that several computer systems, all with their own local memory, can access a common block of *global memory*.

This is an excellent architecture for many fast, large-array data-collection situations. The DPM can be arranged for easy computer access as if it were normal memory, yet be directly accessed by the peripheral I/O system as its own memory. If the design is carefully

Figure 7.31   Block layout of a DPM system.

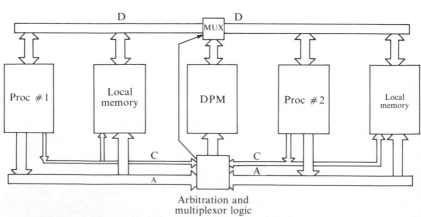

Arbitration and
multiplexor logic

executed, it is often possible to provide totally *transparent access* from both the computer and the peripheral logic. This means that both systems have unrestricted access to the shared memory (as in the Apple VDL). Since the peripheral memory system need not be limited by the computer databus and address bus widths, they can be tailored to suit their applications. The external system's databus can be made as wide as required for external data collection. The external address space can be as big as required, yet can be organised in segments small enough to fit into the limited I/O space of the computer; segments can be selected for direct computer access by a control register in the interface. Very high operation speeds, approaching the limit achievable with digital logic, can be attained by building the peripheral system as a dedicated, *hard-wired*, combinational logic machine rather than as a sequential, programmable machine. Some examples of such data-collection systems are described in Chapter 9.

7.11 A summary and some final remarks

The broad theme in this chapter has been the use of a computer for instrumentation and control; the particular topics have been the methods by which the outside world and the computer can be made to interact.

In summary, there are three classes of such interaction.
(1) Programmed interaction, where the program alone is in control.
(2) Interrupt interaction, where hardware is provided to initiate interaction, but the processor must carry it out.
(3) Direct access, where hardware provides the data transfer between the outside world and the computer memory without any program intervention.

Each of these schemes have fields of application. Speed, power and cost all increase through the three.

Throughout the previous chapters, stress has been placed on the inherent simplicity of the computer's operation. The early microcomputers were simple, open machines and were intended for enthusiasts and tinkerers who wanted to 'get their hands dirty' inside the machine. The machines were extremely easy to interface at the hardware level and were ideally suited to the type of instrumentation applications needed for scientific and engineering applications. Their simple, open structure has dictated their use throughout this book, to explain and demonstrate the principles of computer operation and interfacing. Unfortunately, the commercial pressures of computer marketing, aimed at the lucrative business software market, have forced the popular machines further and further away from the simple, open, 'grey-box' structure of the early machines, with good, full disclosure of their hardware and firmware structure. The newer machines tend to be 'black boxes' whose internal operation and structure is often deliberately obscured from the user. The design of peripheral systems for this type of machine requires considerable knowledge and experience, often more than that of the scientist or engineer who only wants to use the computer as a tool. The newer machines are therefore not very suitable for instrumentation applications unless adequate peripheral support systems are available from their manufacturers. These peripheral systems are usually quite elaborate, so that they can be used in a wide range of applications. This tends to compromise their design, and prohibits them from achieving the full potential power and speed of a well-designed bespoke system.

Yet, not all is lost; simple, open, machine architectures are still available in data-collection/control-system oriented microcomputer systems. These are not as cheap as the large volume, personal microcomputers and are not as well served by mass-produced peripherals and software. The STD bus system is an example; it provides access to a very wide range of excellent data-collection/control modules at moderate cost (refer to Chapter 8).

The scientist or engineer can achieve the benefits of both the cheap, personal computer with its extensive peripheral and software structure, *and* the specialised, control-system modules by interfacing the two together with one of the schemes outlined in this chapter. Alternatively, dedicated instrumentalists can consider designing and building peripheral systems tailored to their own specific needs, and linking them to a suitable, personal microcomputer. (One with extensive documentation on its hardware and firmware systems!) This arrangement turns the personal computer into a cheap, powerful and intelligent peripheral of the data-collection system. Some examples of this approach to computer instrumentation are outlined in Chapter 9.

Exercises

7.1 Design a practical circuit using LSTTL chips, to implement the schematic diagram of Fig. 7.2(*a*) (Box 7.1).

7.2 Design a combinational logic circuit to provide eight separate decode signals for the BBC 1 MHz interface (Fig. 7.5(*b*)), similar to the device select signals of the Apple. Each signal is to correspond to 16 addresses.

7.3 Devise a partial decode circuit for Fig. 7.2(*a*) to expand it to four separate addresses. (You might choose to use a 74LS139 or 74LS155 decoder.)

7.4 Devise a logic circuit to allocate one of the output channels (ports) of Fig. 7.3 as an address expansion register. Use partial decoding of the output bits of this port to generate four ENABLE signals. Arrange the external system architecture so that it expands the interface to four, 8-bit, digital, input channels at each of the 16 device addresses of Fig. 7.3.

Individual 8-bit bus drivers can be used for each of the 64 input channels. Consider the possibility of reducing the number of chips.

7.5 Modify the routine of Example 7.2 in Box 7.2 so that: (*a*) the segment SERVICE is quit as soon as all active status flags are services, or (*b*) the service priority order can be specified by a software array.

7.6 Devise a circuit which uses four of the latch outputs of Fig. 7.3 to strobe two external flip-flops on or off. Arrange that both are turned off by the RESET line going low.

7.7 Using elements of the circuit in Fig. 7.3 and the remarks in Section 7.4.3, design an interface circuit which allows an analog voltage proportional to the product of two 8-bit, digital, output words to be generated.

7.8 Analyse the drive circuit of the 74LS373 bus driver in Fig. 7.15, to understand its behaviour.

7.9 Modify the basic circuit shown in Fig. 7.18(*b*) to accept an EXTERNAL START/STOP signal or separate START and STOP external signals.

7.10 Design an interface using 74LS593 counters, which operates as a programmable frequency synthesiser in the manner discussed in Box 7.5. (If the RCO signal from the counter array is used to reload the counter *and* to toggle a JK flip-flop, the JK flip-flop output will be a precise square wave at half the toggle frequency.)

7.11 Design and build some interfaces for data-collection tasks of your own.

References and further reading

The references in this list are only a small selection from the extensive range of literature available on the topics.

1 *Scientific and Engineering Problem Solving with the Computer*, W. R. Bennett, Prentice-Hall, New Jersey (1976)
A broad, entertaining introduction to software aspects of modelling and simulation.

2 'Going Further', C. A. Pratt, *Byte*, Mar. 1984, pp. 204–8
This article gives references to up-to-date work on computer simulation.

3 *Numerical Methods for Scientists and Engineers*, R. W. Hamming, McGraw-Hill, New York (1962)

4 *Mathematical Methods for Digital Computers*, *Vol. I and II*, A. Ralston & H. S. Wilf, J. Wiley & Sons, New York (1967)

5 *Numerical Analysis*, L. W. Johnson & R. D. Riess, Addison-Wesley, London (1977)
6 *The SY6500 Microprocessor Family – Applications Information* (ibid. Chapter 6)
7 *Assembly Language Programming on the BBC Micro* (ibid. Chapter 6)
8 *Z80 Microprocessor Advanced Interfacing with Applications in Data Communication*, J. C. Nichols, E. A. Nichols & K. R. Musson, H. W. Sams, Indianapolis (1983)
9 *Z80 User's Manual*, J. J. Carr, Reston, Reston, VA (1982)
10 *Z8000 Handbook*, M. L. Moore, Prentice-Hall, Englewood Cliffs, New Jersey (1982)
11 *Apple Interfacing*, J. S. Titus, D. G. Larsen & C. A. Titus, H. W. Sams, Indianapolis (1981)
12 *Interfacing to Microprocessors*, J. C. Cluley, Macmillan, London (1983)
   This reference is Z80 and 8080 oriented.
13 *Microprocessor and Peripheral Handbook* (Intel Data Manual)
14 *NEC 7220 GDC Controller Manual*, NEC Inc., USA
15 *Apple II Monitors Peeled*, Apple Computer Inc., California (1981)
   This manual describes the internal structure of the Apple Monitor firmware operating system. It describes the boot-up procedure and other aspects of operation in detail.
16 Data manuals for linear devices from analog devices, Datel, National, Motorola, Burr-Brown etc.
17 *STD Bus System Manual and Buyers' Guide*, Prolog (USA)

8

Software systems and computer hardware

Preamble

The previous seven chapters have described the internal operation of a microcomputer in a carefully structured manner. The description has been sufficiently detailed to strip away any mystery or magic and to depict a computer as a simple, eminently understandable machine.

The operation of a contemporary computer system involves more than the simple structure described so far, but the reader should now see that sophisticated computers are only an elaboration of simple principles. We could now explore more complicated aspects of microcomputer, minicomputer and mainframe operation at the same level, but they are far too numerous and varied to be explained thus in a single book. These two final chapters will only provide a brief outline of these topics. It is hoped that this will be sufficient to provide adequate insight, or to provide a soft-entry into the extensive literature.

This chapter will describe some of the computer hardware elements and software concepts which turn the simple microprocessor architecture into a powerful computing system. Some of these were mentioned briefly in previous chapters but were neither easily explained nor worthy of a diversionary dissertation at that point. The final chapter will describe the use of microcomputers for instrumentation and data-collection systems.

8.1 Computer peripherals

This section gives a brief outline of the operation of conventional computer peripherals.

8.1.1   Keyboard input

The Apple keyboard input port was discussed in Section 7.3, but no description was given there of the circuitry that detects a keypress and generates the ASCII code for the key.

Keyboards are usually wired in the form of a switch matrix with the keyboard switches at the crossover nodes. Depressing a key connects the matrix wires at that node (Fig. 8.1(a)). In the Apple, keypress detection is carried out by a keyboard-encoder microchip. This chip detects and encodes a keypress by circulating short output pulses through each X-matrix wire one at a time, simultaneously scanning the Y-matrix wires. When the X-scan pulse reaches the pressed key, the Y line pulses at the same time. The coincidence of the X and Y pulses gives a unique address pair to access a look-up table (a ROM array, usually within the chip). This table provides the 7-bit ASCII code for the key.

The keypress code is then latched over to the keyboard input port. The keypress detection also generates a strobe pulse which sets a flip-flop connected to the D_7 status flag. If the REPEAT key is held down, the strobe runs at about 15 Hz. The use of simple switches across the X–Y-matrix limits the number of simultaneous keypresses that can be detected. If more than two are pressed simultaneously, interline feedback causes *phantom keypresses*; this is called 2-key *rollover*. If diodes are used in series with the switches, any number of simultaneous keypresses can be detected; this is called *N-key rollover*. (Read the references for more information on *debounce* and the *2-key rollover* on the Apple keyboard.)

Many computers use interrupt driven keyboard input. Any keypress diverts the processor to a keyboard interrupt service routine within the operating system (see Section 8.7).

Low-cost microcomputers usually have the keyboard interfaced through a simple digital I/O port (usually a PIA or VIA, Section 8.2). Fig. 8.1(b) shows a suitable configuration. If data $00 is loaded into the output port, a keypress can be detected either by programmed interrogation of the input port to detect any low bit there, or by an interrupt signal generated as shown. Either response can divert the processor to a keyboard scanning routine which sets one output bit low at a time, and interrogates the input port until the closed keyswitch is detected.

The BBC microcomputer uses a VIA to strobe the matrix columns continuously; the row outputs are sensed at a VIA input port. Any keypress generates an interrupt.

8.1.2 Video raster character and graphics display

A 625-line television picture requires about 600×600 pixels (picture elements) to be sequentially and repeatedly drawn on the screen by the video raster. The video picture consists of two *interlaced* frames each with half the total picture information, scanned at 50 Hz (Fig. 8.2(a)). A single full-frame TV monochrome picture (no grey scale, 1 bit of data per

Figure 8.1 Keyboard decoding techniques.

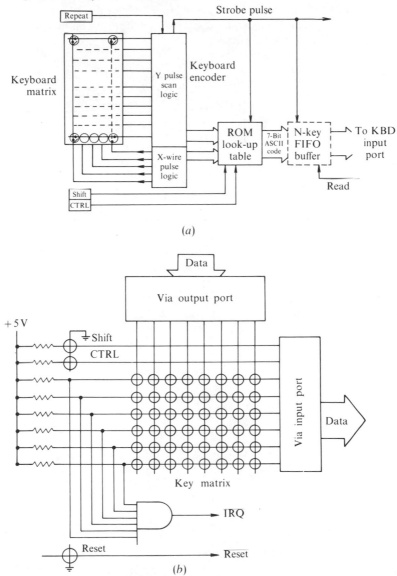

(a)

(b)

pixel), needs about 360 000 bits, about 40 Kbytes, of information to be *bit-mapped* in memory (Fig. 8.2(*b*)).

Data compression techniques can reduce the amount of information needed. A common technique replaces a string of identical bits in the raster display, with the length of the string. This is effective for line diagrams and coloured cartoons, where most of the display is empty or of uniform colour.

The integrated display controllers described briefly in Section 7.9 can be

Figure 8.2   (*a*) Raster display scan and (*b*) the bit-mapping of a display in 8-bit memory.

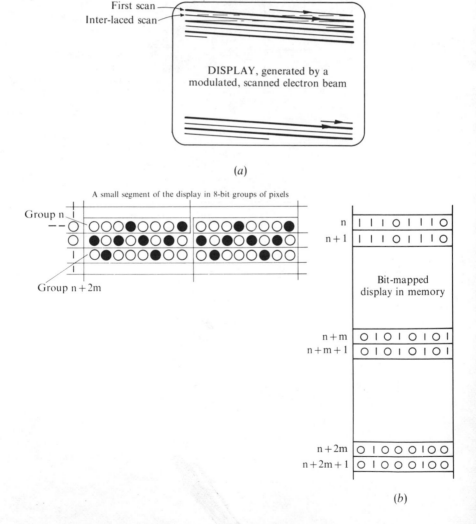

programmed to modify the manner in which the display memory is bit-mapped onto the screen. The NEC 7220 has a 1024×1024 bit-mapped display memory which can be *panned* over the screen by selecting different areas of memory for display on the screen, or can be *zoomed* in or out by changing the memory scan rate with respect to the video scan rate.

It also allows three banks of memory to be used to specify the three separate colour signals required by a three-colour RGB (Red/Green/Blue) monitor. This simplifies the colour display electronics by eliminating the encoder needed for a normal TV colour signal. It also improves the display quality and allows the video scan rate to be determined by the needs of the computer and its memory, not by a television standard. The NEC APC microcomputer uses this chip with three single-bit 1024×1024 memory maps and a 640×475 pixel three-colour display. The display area can be moved about on the larger memory area. The display is not interleaved and is at 41.5 frames per second.

Display controller chips can be used to provide grey-scale monochrome or three-colour displays with individual colour intensity signals. This is achieved by replacing the single-bit pixel memory with a multi-bit word for each pixel, and by adding fast DACs to generate the video intensity for each pixel. (See the references at the end of this chapter.)

Bit-mapping of the display allows absolute flexibility but, if only a small subset of all possible displays is required, coding techniques can greatly reduce memory requirements. The common example of this is for text display, where a small memory array is used to hold pixel bit-maps for each character as a look-up table. The full display can then be defined by a small memory map containing the code for the character to be displayed at each character field of the screen (Fig. 8.3). The memory saving is considerable; the Apple requires only 1 Kbyte for a text screen, but 8 Kbytes for a bit-mapped display.

As the VDL circuitry writes each row across the screen, it fetches the code for each character field and uses the character definition table for the pixel data in the current row.

If the *top-left pointer* can be moved through the text map, and if the map is accessed with rollover from its end to its start, the text in memory can be *scrolled* over the screen. Manipulation of the text map also allows text windows and other features.

The character definition tables are generally held in a small ROM array accessed only by the video generation circuitry. The 256 characters that can be defined by a single-byte code are more than are needed for text. The surplus codes can be used for special symbols or graphical characters. If the character definition table is placed in RAM, software can change the table to provide *limited graphics capability*, without the memory needed for full bit-mapping. Manipulation of the text table can move the graphic characters around the screen to create fast but primitive animation. The character generator table can also be changed to provide alternate text fonts.

8.1.3   Disc I/O hardware

The versatility of a computer can only be realised if programs and data can be moved quickly to and from a bulk-storage medium. The standard medium for microcomputers is the floppy-disc, in one of its several sizes; hard-disc systems are rapidly becoming popular.

This section gives a brief outline of their operating principles; the description is based loosely on the Apple disc system, mainly because extensive, detailed descriptive literature is available for further reading. Single-chip floppy-disc and hard-disc controllers are available and are fully described in computer peripheral data manuals.

The memory space available within a computer is limited by hardware restraints, but any word can be accessed directly by the processor; this is called *Random Access Memory (RAM)*. The storage media used for bulk storage of program code and data *outside* the computer bus structure are invariably accessed serially. An interface is needed to transfer data between

Figure 8.3   (*a*) Character fields, (*b*) text code storage and (*c*) character definition. The definition bit-map for 'A' is shown. It is held in a reserved table in memory, with definitions of all other characters.

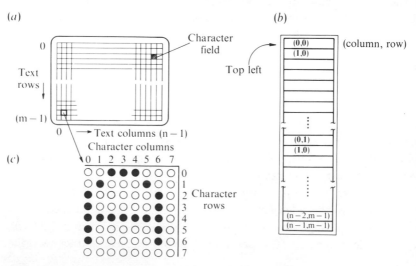

the two disparate systems, and to define its location within the external storage.

A magnetic disc, like a magnetic tape, holds information as a pattern of magnetisation in its 'hard' magnetic coating. Unlike common analog tape recording, and to reduce noise, digital information is stored in binary form by saturating the magnetic material fully in one of two possible orientations (Fig. 8.4). Information is written sequentially as the medium is passed under the write head, and is read back as a series of transition pulses when the magnetic transitions pass below the read head. Pulse detection logic must be used to recover the digital sequence.

Provision must be made to ensure that sequential, recorded data can be reassembled into its original form without a critical dependence on the speed of the medium. Common techniques for providing this during data recording are either to interpose a clock pulse between each data bit pulse (Fig. 8.5(*a*)) or to alter the width of the regular clock pulse so that the data bit is defined by the timing or 'phase' of the subsequent reverse clock transition (Fig. 8.5(*b*)).

The start of the serial data stream on the medium must be defined unambiguously. This is achieved by recording synchronisation bursts to define the start of data words and by adding either mechanical markings to define specific reference locations on the disc (hard-sectored) or by formatting the surface of the disc and writing reference marks and data onto it, before data is recorded (soft-sectored).

Both techniques divide the recording surface into small blocks, called *tracks* or *cylinders* (circular rings of recording) and sectors (angular segments of the tracks). The data is recorded and read sequentially along a track while the disc is spun by a well-regulated drive motor. The track is

Figure 8.4   Digital magnetic storage.

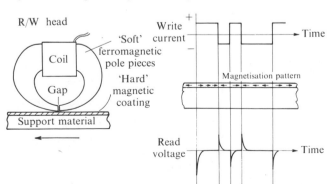

selected by a motor (usually a rotary or linear stepping motor, see Section 9.1) which can move the recording head radially across the tracks (Fig. 8.6(b)).

Before a new disc can be used, it must have all of its sectors defined on the surface; this is called *formatting*. Reserved codes are recorded on the disc to mark and label each sector. These codes cannot be used in the user data field lest they be misinterpreted as markers. The computer data stream must therefore be processed to avoid reserved codes, and encrypted before being recorded. The data stream read from the disc must be decrypted. Error detection is provided by adding some redundant information. This is usually a *checksum* word which, for 8-bit data, is just the sum of all the data bytes modulo-256. Fig. 8.6(a) shows the code marking and sector organisation in the Apple DOS (Disc Operating System).

The encrypted, serially accessed data on the disc surface is not suitable for direct program access. Data access is usually hidden behind the DOS software which takes care of the formatting and encryption. The lowest level of direct user access is via the system routines which allow a track/

Figure 8.5 Bit patterns and synchronisation schemes. (*a*) Interposed data pulse and (*b*) transition timing (phase) encoding.

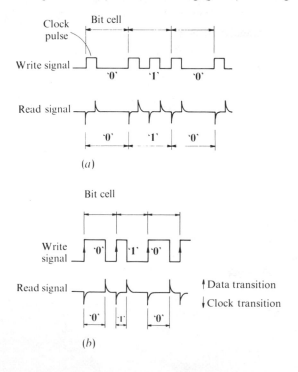

sector address to be defined, and instruct the system to perform a read or write between the data field thus defined and a block of *buffer* memory within the computer. (This is the RWTS routine of DOS.) When the contents of the user data field in the sector have been transferred to the buffer area, the data can be written, read or modified by normal memory access. When data on another track/sector address is required, the system software must restore the buffer contents to the correct place on the disc and must fetch the next segment into the buffer. Thus, a user file consists of several sectors, linked by a track/sector list. A file must be OPENed before it can be used and should be CLOSEd when the program is finished with it; the first command assigns the file to a buffer, the latter forces the DOS to restore the buffer to the disc. If several buffers are available within the operating system, several files can be held open simultaneously for reading, writing, transfers or modification.

DOSs usually provide a range of routines to allow the DOS software to be booted-in at power-up, to set directories of named files (along with a list of the track/sectors used by each), and to provide a range of simple high-level commands to establish, load and retrieve various types of user files.

8.1.4 Communication hardware

The concepts of data transactions and handshakes described in Chapter 4 represent the basis of all digital communication systems. Within

Figure 8.6 (*a*) The Apple DOS 'soft-sectored' track format (one sector shown) and (*b*) the track-sector layout on a disc.

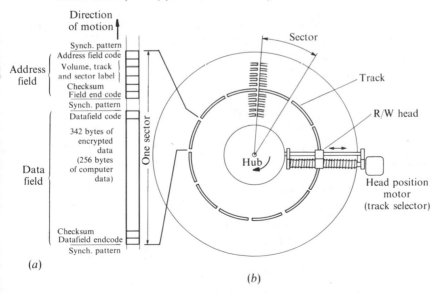

(*a*)

(*b*)

a computer, the transaction protocol, the timing and the data format are completely defined by the *synchronous* hardware of its *closed* communication system. In the more general case of an *open* system linking remote asynchronous systems, protocols must be defined for all aspects of communication.

A wide variety of protocols are defined as digital communication standards; new or modified standards appear regularly as the available technology changes. It is pointless in a book such as this to describe them all. Instead, the general requirements of a communication system, and some features of the standards relevant to data-collection applications, will be described.

The procedure involved in a telephone call demonstrates the requirements of an open communication system. When a caller wishes to communicate with another subscriber (the *callee*), access to the system is made by lifting the handset. The system responds with a dial-tone. This simple *handshake* allows the caller to move on to the next phase of the communication.

The caller sends information to define the callee required. The system establishes the link using coded signals to indicate the system status. If the connection can be made, it issues a *ring* signal. If the callee answers, the link is established. The next phase involves both caller and callee in a protocol for mutual identification. The callee then enters the next phase by indicating whether information is to be given or to be received, and the content (and format) of the data. The data transaction can then take place. If the data is critical, some form of check will be made that it has been received correctly.

The caller and callee use the universal conversational protocol of common language; its subtle inferences allow changes of phase and state without conscious effort.

Machine communication cannot be based on subtle inferences, it must be fully defined by a nested set of protocol tiers. The lowest tier defines the binary signals used within the system. The next tier defines the timing and the handshake sequence. The topmost tiers define the procedures used to change the transaction phase, the format of messages passed over the link and the technique used to check data.

Within a microcomputer bus system, these three tiers correspond respectively to the bus signal levels, the control bus timing and the program code. They are only applicable within the tightly specified environment of the computer and its peripherals. The absence of standards for computer bus protocols requires that computer–computer communication be made via an alternate, universal, communication system.

In the old electromechanical *Teletype* system, the binary signal at the lowest tier of the communication protocol was the presence or absence of a current of about 20 mA. The contemporary protocol is usually the EIA specification of 3 V (and above) for a logic **0**, − 3 V (and below) for a logic **1**. These signal levels are referred to the local ground. Any ground noise between the sender and receiver will degrade the signal levels, hence this protocol can only be used over short distances. Long-distance communication in noisy environments requires a balanced line link, with two wires for each binary signal. The two lines are driven with complementary current or voltage signals; a difference detector is used at the receiver to filter the information signal from *common mode* ground noise. Standard line driver/receiver chips are available to handle fast logic signals over moderate distances with many volts of ground noise.

Lightning strikes, or large electrical surges, can induce thousands of volts between the ends of a communication line. Transformer coupling (for a.c. signals), fibre optic communication or optical isolator chips cope with this.

The next tier of protocol defines the timing and control functions; it defines the method for designating the caller and callee (the master and the slave), the direction of data flow and the data transaction handshake sequence.

The *RS232 Standard* communication system provides 18 dedicated signal lines to handle all aspects of the control and handshake protocol. (It uses EIA (Electron Industries Association) signal levels on all lines.) Data transmission is made on a single data line but separate lines are provided for either direction.

In the *asynchronous* mode, the sender starts a data string by sending one or two *start bits* followed by seven or eight bits of data; the string is ended by a *stop bit*. The start bit causes the receiver to begin sampling the serial bit-string at an agreed rate.

The short data string was needed in earlier days of communication when stable crystal controlled clocks were not freely available. The bit-rate of the data transmission is defined in the protocol, but the poor long-term stability of economical timing circuits restricted the length of a bit-string that could be transmitted reliably. Many bit-rates are allowed from 100–20 000 b.p.s. (bits per second). The data format is not specified by the standard but it is generally used for text transmission using ASCII code. RS232 also defines a *synchronous mode*, data transmission protocol. Four of the control lines are dedicated to timing signals; the sender or receiver can transmit clock signals to synchronise the transmission or receiver data rate. Provided that both the sender and receiver know the number of bits in a data string, it can be of any reasonable length.

Further control lines provide data transaction handshakes in either direction, and other general control functions. The master/slave relationship is not negotiable within the control protocol; it is defined by the wiring at the terminals.

The IEEE-488 standard bus, which began as the Hewlett-Packard Interface Bus (HPIB), was developed originally as a communication system for minicomputers to control and collect data from instruments such as digital voltmeters and frequency meters. Three types of device are defined: controllers, talkers (senders) and listeners (receivers). The bus contains 16 lines all at TTL logic levels. Eight of these are a bidirectional databus, the other eight are dedicated control lines. There is no address bus, but the controller device can force the system into its addressing state by signalling on a dedicated control line and by depositing the address of the required device on the databus. The controller can thus force the system into the addressing state and can then nominate one talker and any combination of listeners for subsequent data transactions.

Up to 15 separate devices can be connected (subsequent enhancements of the standard allow expansion). Data transactions are asynchronous under a full handshake protocol; data rates up to 2 Mbyte/second can be achieved.

Single-chip, dedicated processors are available to interface the IEEE-488 bus to a microcomputer bus.

Economical long-distance communications require a minimum number of signal lines. Time-domain bidirectional serial transmission must be made on a single line. The telephone system uses a single subscriber line and a common ground return for ringing the receiver's bell, carrying addressing code, spoken information transmission and to provide power for the handsets. Simple d.c. logic signals cannot be used; information must be represented by a.c. signals. The common technique for binary signals uses either two different frequencies (Frequency-Shift Keying (FSK)), phase changes of a steady tone (phase modulation) or a combination of amplitude and phase change (quadrature amplitude modulation). (Combination of these allows several binary signals to be transmitted simultaneously.)

Binary logic signals must be converted to the frequency modulated line signal by a *modem* (MOdulator/DEModulator) at the transmitter and must be demodulated at the receiver. FSK modulation allows several distinct frequencies to be transmitted simultaneously on a single line. By defining two pairs of different frequencies for data sent in either direction, it is possible to transmit and receive simultaneously on a single line. (If the frequencies are chosen carefully to avoid harmonic interference, several signals can be multiplexed on a single line.)

The multiple control lines of RS232 and IEEE-488 allow the communication process to be moved through its necessary phases by the hardware. The telephone system uses the single subscriber line for all phases of the communication process, the protocol must therefore define special codes or formats that allow changes of phase and state to be defined unambiguously.

Hewlett-Packard have developed a communication link for small computers and instruments called the *HPIL* (*Hewlett-Packard Interface Loop*). This loop uses a single twisted-pair line for all addressing, control and data transmission functions. The signal protocol provides separation of address, control and data states.

Each device in the loop accepts input on one twisted pair and sends or passes data through a second output pair. Information is passed by signal transitions rather than by levels; this allows a.c. connection through transformers and eliminates both d.c. ground loops and common mode voltage transients. Data transmission rates are in the order of Kbytes per second. The system is very well thought out and will no doubt soon become an IEEE standard for small systems.

The topmost tiers of the communication protocol define the structure and format of the data stream. Fig. 8.6 demonstrated this for disc storage, showing how a *coded header* is used to mark and identify each sector and track, to provide information about the data, and to add *checksums* to confirm the integrity of the header and the data.

In a closed system such as a disc drive, the coded header need only provide a limited amount of information. The system hardware and software provide the rest of the protocol, to specify data field lengths, to link data fields into files, to name them and to define the data format. In an open communication system, however, there is seldom any hardware to organise the extent and format of the data. All must be done by implicit agreement between the terminals or by a further tier of protocol that uses a prologue sent before the data stream, to define its content and format.

A good example is the transmission of packets of data over a *ring network*. On a ring, the data is transmitted in one direction from terminal to terminal until it reaches its destination. The data packet must obviously contain both the caller's and the callee's address. The header should therefore contain two addresses and some information about the quantity and format of data in the packet. Redundant information for error detection (and possibly correction) would usually be inserted. The system must be able to distinguish header information from true data. This usually requires reserved codes or control patterns to mark the header.

The caller and callee are able, by arrangement, to add further layers to the protocol structure. The first words of the transmitted data could, for example, be used to define the subsequent data as program code, raw binary data, text or whatever.

The concept of a header block is extremely flexible and powerful; it is applied universally in all types of data transmission and storage systems.

8.2   Microprocessors and support chips

Every microprocessor family is supported by an extensive range of VLSI peripheral chips. They provide simple, efficient access to the computer busses for interfacing standard peripherals and simple data-collection control systems, but their closed structure limits their use in many unconventional instrumentation applications such as those to be described in Chapter 9.

8.2.1   6502 support systems

The 8-bit 6502 is supported by a moderate range of chips. The 6520/1 Peripheral Interface Adapters (PIA) provide a pair of bidirectional I/O ports and two control registers that can be programmed to provide data transaction handshakes or automatic interrupt generation from the I/O ports.

The more elaborate 6522 Versatile Interface Adapter (VIA) provides a wide range of interface functions. It has 16 internal registers, and allows programmed access to I/O ports, two 16-bit timer/counters, an 8-bit shift register and interrupt status/control registers (see Section 7.7).

The timer/counters can interact and can be programmed to time or count external events, to generate timed interrupts or to control an internal shift register for parallel-to-serial or serial-to-parallel conversion (see Sections 2.2 and 8.1.4). The VIA can be used for peripheral I/O functions such as keyboard input, serial I/O, parallel printer output etc. It can be used effectively for simple data collection in programmed or interrupt interaction schemes.

The 6545 Cathode Ray Tube Controller (CRTC) chip provides the VDL functions discussed in Sections 8.1.2 and 7.9. It only allows text and limited graphics display.

The 6551 Asynchronous Communication Interface Adapter (ACIA) provides all of the logic hardware required for interfacing an asynchronous serial communication system. (See Section 8.1.4.)

The 6591 Floppy-Disc Controller (FDC) provides all of the peripheral functions and control protocol for a floppy-disc drive. It provides interrupt interaction for block data transfers between disc and memory as a background processor task.

A support chip (6530/2) is also available to provide all of the functions needed to convert a 6502 into a self-contained process control microcomputer for control applications and simple data-collection systems. It contains a small block of RAM, some ROM space (programmed during manufacture by a custom mask) and a simple I/O structure.

Self-contained microcomputer systems can also be built from specialised single-chip microcomputers such as the 6500 and 6511. These have a 6502 processor, a powerful VIA I/O structure and memory in a single package. The standard configuration contains a small block of RAM and a block of mask-programmed ROM which limits them to mass-produced controllers. Emulator chips, intended for development of the single-chip system, have no internal ROM and can use external memory. They can be used for efficient one-off or low-volume controllers, or for smart peripheral systems.

8.2.2   Z80 support chips

The 8080/Z80 family of processors is supported by an impressive range of chips from Intel and Zilog. Interrupt and DMA operation is well supported; internal interrupt daisy-chain logic is provided to allow powerful, vectored interrupt.

Parallel Input/Output (PIO) chips provide parallel I/O ports and internal control registers. Simple, or handshake, data transactions can be specified. They can be programmed to generate interrupts from a variety of external logic conditions and to provide the interrupt vectors required by the Z80's IM2 interrupt response (Section 7.7).

Counter/Timer Chips (CTC) contain multiple 8-bit counter/timer functions with individually programmed modes, such as prescaling and automatic preload followed by down-count from the prescaler or from external events. Interrupts can be programmed to occur on specific counter states.

Serial I/O interface (SIO) chips are general-purpose serial–parallel and parallel–serial drivers for communication links. They support all UART (User Asynchronous Receive and Transmit) and USART (User

Synchronous and Asynchronous Receive and Transmit) functions. Parity generation and general serial packing/unpacking are handled internally.

Direct Memory Access (DMA) controller chips (Section 7.8, Box 7.7) are oriented to block data transfers but can be used for simple DMA access with instrumentation systems.

Dual-Ported Memory (DPM) controllers and FIFO (First In, First Out) 8-bit word buffers are also available as bus-oriented single chips.

Several single-chip microcomputers are available with similar specifications to those based on the 6502.

8.2.3   Other 8-bit processor support chips

The Motorola 6800 series of microprocessors provides an extensive range of devices for control and data-collection applications. Special note should be taken of the analog I/O provisions, and of the low-power CMOS range of chips for battery-powered applications. Extensive literature is provided by the manufacturers.

8.2.4   Some 16-bit microprocessors and support chips

Most 16-bit microprocessors are 'system oriented' and use powerful support chips for large system and memory management (Section 8.4).

The 65SC816 is a 16-bit expansion of the 6502, with a 24-bit address bus and a 16-bit internal databus. The top eight address lines and the two data bytes are all multiplexed through eight pins. A single input switches it to emulate the 6502.

The Intel 8086 is upwardly compatible with the 8080 (and Z80) family of processors. It multiplexes data and address information on 16 address/data lines; during the first phase of a bus cycle, the 16 pins provide address information. A set of five support chips (one 8288 bus controller, two 8282 latches and two 8286 transceivers) are needed to create the full 20-bit address bus (see Section 8.4.4) and the 16-bit databus.

The 8289 Bus Arbiter support chip allows multiple processors to operate on a single bus system. It handles the bus master–master handshake described in Chapter 4, Box 4.4.

The 8087 arithmetic co-processor chip can be connected in parallel with the 8086. This enhances the instruction set with extra opcodes recognised by the 8087 alone. When these codes are passed to the processors, the 8087 operates the control pins of the 8086 to take over instruction execution. The 8087 provides fast numeric data processing on integer and floating-point numbers; typical execution speeds are up to 100 times as fast. The 8087 achieves this performance by internal microprograms for numeric instruction execution sequences performed within internal registers (Chapter 5).

The 8088 8-bit bus version of the 8086 uses an 8-bit databus and time-multiplexes 16-bit data transactions.

The 8086 and 8088 can be interfaced in much the same manner as the 8080/Z80; bus transaction protocols are very similar. They can use many of the 8-bit 8085 microprocessor support chips for I/O operations. The 8259 programmable interrupt controller provides an 8-level (expansible to 64 levels) interrupt request structure.

The 8207 Dynamic Ram Controller provides all of the logic needed to control large memory arrays made from 16K, 64K or 256K dynamic RAMs. It also allows DPM access.

The Zilog Z8000 was designed with no compromises and is incompatible with the Z80. It has 16 multiplexed address/data lines, hence needs some latches etc. for the bus system. There are 16, 16-bit general-purpose registers which can be used for 8-, 16- or 32-bit words. The address bus is 23 bits wide with a 7-bit segment pointer (see Section 8.4).

Support chips include a DMA controller, a communication controller, a peripheral controller, a VIA and a FIFO I/O interface unit.

It is very well designed and is oriented to large systems, but it has not gained the popularity that the 8086 has achieved with design compromises allowing software support by upward compatibility.

The Motorola 68000 is a very powerful 32-bit processor with a 16-bit databus and a 24-bit address bus; the chip is in a 64-pin package to avoid the need for the address and data busses to be multiplexed. It implements instruction pipelining; while the processor is executing an instruction, and if the external busses are idle, the subsequent instruction is brought into a *prefetch queue* ready for execution. An extensive range of all the normal support chips is available. Bus transaction sequences are rather like the 8-bit 6800 family; interfacing is quite straightforward.

8.3   Memory systems and programmable logic

This section describes the fundamentals of programmable memories for microcomputers, and programmable logic arrays.

8.3.1   Random access memory

The name *Random Access Memory* (*RAM*) is a hangover from first- and second-generation computers where almost all memory media could be written and read, and a distinction was made between memory that was accessed serially (tape, drum, disc etc.) or accessed randomly. The name is now used quaintly to distinguish random access R/W memory from random access memory that can only be read, *Read Only Memory* (*ROM*). There are, of course, some types which do not properly fit in either of these classes. This section will outline the operating principles of the common types of R/W RAM.

The basic elements used in electronic binary memory were introduced in Sections 2.1 and 2.5. Address decoding for large memory arrays was described in Chapter 4, Box 4.2. We will now take a brief look at the electronic detail within the memory chips.

Some additional logic is needed to connect the simple MOS memory cell shown in Fig. 2.21 into an array. Two extra transistors are needed to select a single cell from the array and to connect it to common data lines. The full bit cell is shown in Fig. 8.7(*a*). These cells are arranged in a matrix as shown in Fig. 8.7(*b*).

When one of the row select lines is activated, all bit cells of that row are connected to the data column lines. The complementary I/O data lines for each cell in the row are connected to the R/W logic block. Column-

Figure 8.7   (*a*) MOS memory bit cell and (*b*) an array of cells.

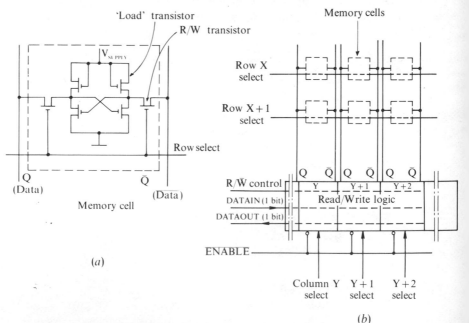

selection logic allows one column line pair to be connected through to the 1-bit data R/W logic.

A single cell is identified when its row and column addresses are presented to the address decode logic; the R/W logic signal determines the action to be performed. Some form of CHIP ENABLE control input is invariably provided; the R/W logic will not respond unless the chip is enabled. This allows many identical chips to be connected in parallel as a big memory array (see Box 4.2).

Static RAM array chips are available up to $32K \times 8$ bits and can be connected into very large arrays. Typical access times are 150–300 ns. Multi-bit word memory arrays consist of multiple arrays as shown in Fig. 8.8, with common address and control lines, but with separate data busses, one pair to each bit of the word.

The general arrangement of Fig. 8.8 is used in nearly all single-chip memories. The different types of semiconductor memories vary only in the bit cell circuit.

Bipolar memory cells are very fast but they are quite expensive and they require a lot of power. The bit cell is shown in Fig. 8.9. While a row is inactive, its select line is held at ground reference potential, and the cross-connected flip-flop will hold its state indefinitely. (This is *static memory*.) When the row select line voltage is raised (to about $V_{cc}/2$), the emitter current from the active transistor of each cell in the row is diverted through

Figure 8.8   The general arrangement of a 1-bit memory matrix chip.

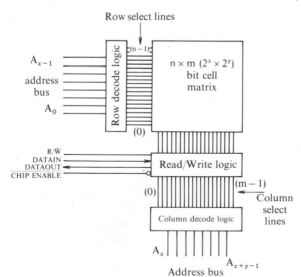

the alternative emitter to the column lines and thence to the R/W logic. Column select logic allows any cell to be read or, for a write operation, allows one flip-flop pair to be forced into either of its stable states. This kind of memory is available in arrays containing up to about 4K cells arranged in configurations of 1-, 4- or 9-bit words. Access time is typically 50 ns. ECL bipolar memory is also available with access times of about 10–25 ns.

Fast, high-density CMOS static RAM modules are also available. The IDT7M656 module (Integrated Device Tech.) has 256 Kbits which can be configured as 16-, 8- or 4-bit words; access time is about 60 ns.

Great economy of space, cost and power can be achieved by replacing the *static memory* cells in Figs. 8.7 and 8.9 with *dynamic memory* cells. The six-transistor MOS cell of Fig. 8.7 can be replaced with a four-transistor cell if the two load transistors are eliminated and their role is taken by the two read transistors (Fig. 8.10). While inactive, the flip-flop is deenergised but holds its state by the charge held in the gate capacitance. This charge leaks away in a few milliseconds (hence the name dynamic) but all of the flip-flops

Figure 8.9   The bipolar memory cell.

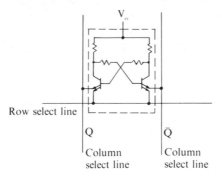

Figure 8.10   The four-transistor MOS cell.

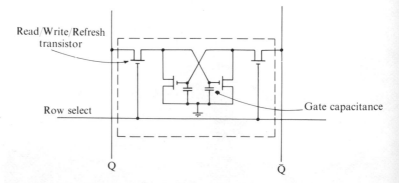

in a row can be *refreshed* (the charges restored) by selecting the row for a normal or *dummy read* of any cell in the row. Obviously, all rows of the memory must be accessed repeatedly to refresh the memory before the gate charges have dropped below a safe, detectable level.

The single-transistor dynamic MOS cell (Fig. 8.11) is the simplest possible memory cell. The row select line turns on the single transistor of each cell in the row, thereby connecting each integrated capacitor to its column line. The tiny charge sent from each capacitor is detected by sensitive charge amplifiers on each column line, and the control logic then recharges each capacitor from individual column *write* amplifiers, before the row is deselected. The *destructive-read* must be followed by the rewrite cycle for memory refresh, but an alternative, fast read/modify/write cycle can be executed by changing the data between the read and write phases. (This facility is seldom used in a computer but is available for fast hard-wired logic (see Chapter 9).) The compact geometry of this simple cell has provided the cheap 64 Kbit chip; 256 Kbit chips are available and the 1 Mbit chip is very close to full commercial implementation.

The 16K and 64K chips would normally require 14 and 16 address lines respectively to define each cell; to minimise the external lines, the address is multiplexed into the chip as two separately latched 7-bit or 8-bit words. The control sequence required for accessing data in the chip is outlined below.

(1) The 8 bits of the row address are presented to the address inputs and the *Row Address Strobe* (*RAS*) input is then forced low. This latches the row address into the row selection logic.

Figure 8.11   The single-transistor memory cell.

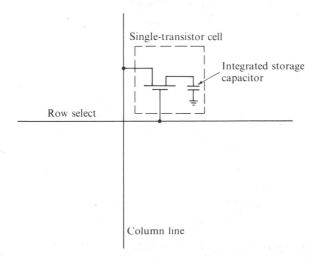

(2) The 8 bits of the column address are then presented to the (same) address inputs and *Column Address Strobe* (*CAS*) input is then forced low. This latches the column into the column select logic and *enables* the chip; it then responds to the *Write Enable* (*WE*) input.

(3) If the WE input is held high during the sequence, the addressed memory cell's data is latched to the single data output of the chip (READ mode). If the WE input is held low, the data at the single data input is stored at the addressed cell (WRITE mode). If the WE input is pulled low at the correct time during the sequence, the previous contents are latched at the output, and the data at the input pin is stored (READ/MODIFY/WRITE mode).

If Step (2) above is omitted, the addressed row is only refreshed (READ/ REWRITE of each cell in the row) and the chip does not carry out any external read or write transactions. The CAS thus behaves like an ENABLE signal. A full array of chips can therefore be refreshed together while only one is enabled to be read or written; this allows powerful and simple memory organisation.

The strobed address technique allows an *extended page mode* access to the chip. Once the row address has been strobed, the column address can be changed and strobed repeatedly and a read, write or read/modify/write of any cells of the row can be made. The time available for this is limited, of course, by the need to refresh all rows.

Some manufacturers provide the facility for *hidden refresh* cycles. After the row and column addresses have been latched, and provided that the CAS input is held low and that the chip is held in READ mode, the address data at the inputs can be changed and the RAS can be pulled low repeatedly to refresh all selected rows, without affecting the output data read from the chip.

The earlier standard 16K Dynamic RAM (DRAM) chip used 7-bit row and column address words. Thus, a microprocessor such as the Z80 with internal memory refresh only generates a 7-bit memory refresh address. Some manufacturers supply modified 64K chips with 7-bit refresh addressing. This is very easily implemented. In moving from 16K to 64K, some manufacturers did not start again from scratch; they reduced the 16K memory layout as small as possible and crammed four of the arrays onto a slightly bigger chip with modified drive circuitry. The bottom 7 bits of row and column are configured as before, but the top 2 bits select the 16K bank. The bottom 7 bits of the row address thus access all cells, 4 rows at a time.

DRAM arrays are usually organised as single-bit memory. Access times can be as low as 150 ns, but a full read or write requires about 250 ns and a read/modify/write cycle about 350 ns, all under ideal conditions.

8.3.2   Read only memory

Second- and third-generation computers used diode matrices for the short *bootstrap* routines needed to get the machine running from a cold start (Sections 6.6 and 7.6). The matrix contained a row for each address of the routine in memory, and each column of the matrix was connected to the respective bit of the databus through suitable buffers. Diodes could be connected between the rows and columns (soldered or plugged in) as shown in Fig. 8.12 such that each row (memory address) selected would impose its hardwired dataword onto the databus. This created a simple, permanent *Read Only Memory* (*ROM*).

MOSFET VLSI microcircuits have made it possible to produce vast arrays of simple transistor circuits to perform the diode function of Fig. 8.12. Fig. 8.13(*a*) shows a segment of a typical circuit for such a *mask-programmable ROM*. The data pattern held in the chip is determined during manufacture by modifying the diffusion or implantation mask so that *functional* transistors are created only at the required matrix intersections.

The technique can provide a wide range of word-sizes and address ranges. The Intel 2316, or its equivalent, is popular for microcomputers. It has a 128×128 MOSFET matrix organised as a $2K \times 8$-bit memory array. Several of these usually hold all of the system firmware of the machine. Fig.

Figure 8.12   Diode matrix ROM.

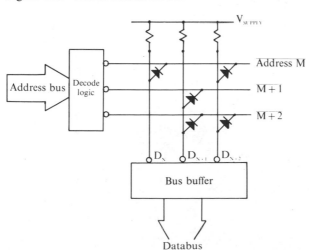

8.13(b) depicts the internal block layout. The masks are very expensive and are not suitable for prototype or one-off designs. Adaptations of the technique provide simple user *Programmable ROM* (*PROM*).

The fusible-link **PROM** uses bipolar transistor technology; as its name suggests, it can only be written once. The chip is programmed on a *PROM*

Figure 8.13   (a) The MOSFET mask-programmed ROM matrix and (b) the 2316 layout.

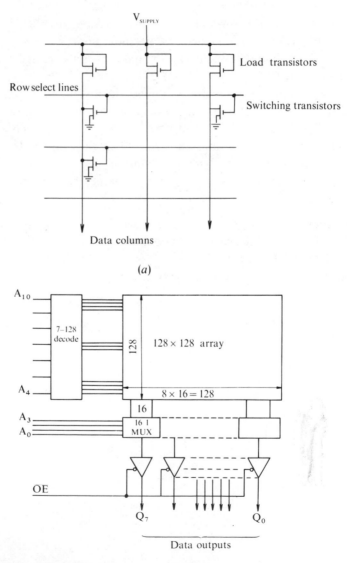

burner by controlled burnout of fuse links in series with the diodes or transistors of the matrix. They are now used only where high-speed ROM is needed; Schottky technology provides access times of about 40 ns for arrays of up to 2 Kbytes.

Subsequent development of convenient *Erasable PROM* (*EPROM*) chips has made the use of ROM extremely convenient for one-off and small-volume designs. There are several types of EPROM array. They use avalanche breakdown or the tunnelling effect to inject charge into the floating gate of a FET (Field-Effect Transistor) (Fig. 8.14(*a*)). (The floating gate is a conductive layer electrically isolated from the rest of the transistor and the substrate.) By controlling the bias voltage between the floating-gate/substrate junction during programming, charge can be injected into the floating gate; when the voltage is reduced, the charge is trapped there. Higher memory density is achieved by building the transistor with a 'stacked' floating gate (Fig. 8.14(*b*)).

The stored charges in the floating gates of these elements dissipate very slowly and are almost certain to outlast the life of their computer. The stored charge can, however, be raised out of its electrostatic potential well by interaction with ultra-violet radiation. Hence the memory can be erased and rewritten at will. It is often called *Ultra-Violet-Erasable PROM* (*UVEPROM*). The standard 2764 chip provides an 8K × 8 array (8 Kbytes), the 27128 provides 16 Kbytes and the 27256 provides 32 Kbytes of ROM memory. Access times are quite long at about 300–700 ns. Erase times are about an hour under moderate ultra-violet sources.

Electrically-alterable EPROM (EEPROM) chips are available. Their construction allows the stored charge in the floating gate to be removed by

Figure 8.14   Schematic layout of EPROM bit cells. (*a*) Separate selection and storage transistors. (*b*) The 'stacked' single-transistor structure.

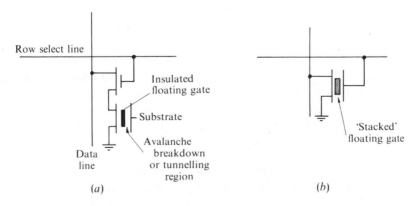

applying suitable voltages to the storage cell elements. (See the NEC NMOS and the Intel 28xx series devices.) Access times are about 200–500 ns. Erasure times are seconds.

The diode or transistor matrix is essentially a code conversion device or a look-up table. Each address pattern elicits a specific, stored data pattern. This property can be put to many other uses than storage of computer instructions.

An obvious application is in the conversion of coded data, for example, conversion of keyboard X–Y-matrix code to 7-bit ASCII code, from binary input to BCD output, for a square root, sine, log etc. look-up table or to convert from GRAY code to binary code. (The GRAY code is favoured for linear and rotational position detectors. Each adjacent position's code only allows one bit to change, hence changes are unambiguous and cannot be effected by *skew* or misalignment of a multi-bit position sensing encoder.)

Another example is the conversion of a BCD code to the specific signal patterns required for character display on seven-segment numeric displays and video screens (see Fig. 8.3 for the table for the letter A). The ROM address is derived from the ASCII code for the character and the 3-bit address defining the row being scanned. The 7 bits corresponding to each row of the display character would be the ROM output.

Industrial processes often require a fixed-sequence controller to generate a timed pattern of signals. A ROM can be used in this application. The control processor simply sends a sequence of addresses to the ROM; the multi-bit ROM outputs represent the output signals required for the process.

The code conversion scheme can also be applied to a combinational logic function generator. Consider the four-input, two-output logic function defined by the boolean expressions below.

$$Q_0 = \bar{A} \cdot \bar{B} \cdot \bar{C} \cdot \bar{D} + A \cdot B \cdot C + \bar{A} \cdot \bar{B} \cdot C \cdot D$$
$$Q_1 = A + B \cdot C$$

These expressions can be generated by a simple ROM. With four independent, and two dependent, variables, the ROM look-up table must be equivalent to a 16×2 array with 32 nodes.

The equations above only have five combinational terms, but when expanded to the full set of product terms they need the 14 matrix terms shown in the table of Fig. 8.15.

The redundancy in the example above arises because all 16 possible product terms must be present in the look-up table. This redundancy is eliminated by devices called Programmable Logic Arrays (PLAs or PALs), described in the next section.

8.3.3 Programmable sequencers and logic arrays

The ROM array's inefficiency in generating sequential or combinational logic expressions is overcome by PLAs. These are similar to the ROM, with full arrays of fusible diodes or transistors arranged as two-dimensional matrices. They are, however, organised in two-tiered structures as shown in Fig. 8.16. The simple boolean product terms of the equations (with one or more variables), are generated in the top section, and the sums of products are generated from the products, in the lower section. Fig. 8.16 shows the residual diode matrix which generates the expressions used for Fig. 8.15.

Commercial PLAs are available with quite large matrix arrays. Simple 20-pin DIL chips provide about 16 inputs driving 32 product lines, and about 8 outputs. Larger arrays are available in 24-pin chips. Variations on the simple structure shown provide feedback from some outputs as further inputs, or provide clocked flip-flop outputs which can be fed back to create programmable sequence generator arrays. The devices are extremely versatile; they can be used to replace large blocks of combinational or sequential logic circuits with a single chip.

This type of logic array is universally used to implement the microprogram sequence in a microprocessor. During the development of

Figure 8.15   Generating combinational logic functions with a ROM array.

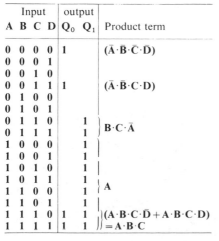
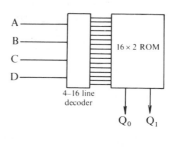

Input				output		Product term
A	B	C	D	Q_0	Q_1	
0	0	0	0	1		$(\bar{A}\cdot\bar{B}\cdot\bar{C}\cdot\bar{D})$
0	0	0	1			
0	0	1	0			
0	0	1	1	1		$(\bar{A}\cdot\bar{B}\cdot C\cdot D)$
0	1	0	0			
0	1	0	1			
0	1	1	0		1	
0	1	1	1		1	$B\cdot C\cdot\bar{A}$
1	0	0	0		1	
1	0	0	1		1	
1	0	1	0		1	
1	0	1	1		1	
1	1	0	0		1	A
1	1	0	1		1	
1	1	1	0	1	1	$(A\cdot B\cdot C\cdot\bar{D}+A\cdot B\cdot C\cdot D)$
1	1	1	1	1	1	$=A\cdot B\cdot C$

the instruction set, changes can be made by fairly simple modification of the array logic. Some microprocessor manufacturers suggest that their chips can be microprogrammed to suit the user.

8.4   Virtual memory, operating systems and memory management

8.4.1   Virtual memory techniques

This section gives a brief outline of the principles used in small computers, to allow an expanded *virtual memory* to be fitted into the limited *physical memory*. The related *memory management systems* used in large machines are also mentioned.

The enable input of memory chips allows them to be connected in large arrays with common address and data busses, yet to be distributed logically throughout the physical memory space. The same principle can be used to *overlay* multiple memory blocks on one physical area.

The BBC microcomputer offers a simple example. The memory space from \$8000 to \$BFFF (Fig. 7.5(*a*)) is occupied by 16 Kbytes of ROM. This is normally used by the BASIC ROM chip (a 27128 16K × 8-bit array), but

Figure 8.16   PLA matrix function generator equivalent to Fig. 8.15.

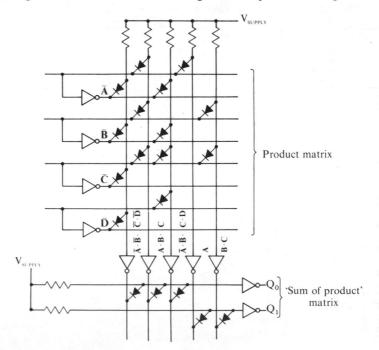

sockets for up to three more 16 Kbyte ROM chips are provided in the machine. The I/O address space $FE40–4F contains control registers which selectively enable one of these four ROM arrays or *memory banks*. This provides a *virtual* space of 64 Kbytes to be selected in 16 Kbyte blocks under program control, to occupy the 16K *physical* address space.

A similar scheme is used in the Apple microcomputer. The 12 Kbyte address block from $D000 to $FFFF is occupied by ROM chips (6 × 2316). The active-high enable lines of all six chips are tied together and held high by a 1 kΩ resistor. The common line is presented to all peripheral slots of the computer. A device in any slot can disable all of the resident ROM arrays by pulling this line low and can then enable its own 12K of ROM to occupy the same physical address space. This provision was made so that the user could insert a *language card* and choose either the APPLESOFT (and MONITOR) ROM, or the earlier INTEGER BASIC ROM (with its slightly different MONITOR). When RAM memory is used in this language card, the system software for any other language can be loaded from disc.

The simplest way to provide a lot of expansion memory would seem to be to add complete 64K banks of extra memory and to select one bank at a time. This poses a small problem, since the program vanishes when the bank is changed. The solution is to provide only segment or quadrant switching, and to keep the program in a permanent segment. Disconnection of I/O devices is somewhat impractical, thus a memory-mapped I/O machine such as the Apple or BBC can only allow memory segment overlays outside the I/O area.

No provision is made for expansion in the BBC machine; the two free pages Fred and Jim can, however, be used for a single page overlay of expansion memory, as described in Chapter 7, Fig. 7.4(*b*). The contents of the page pointer augment the bottom 8 bits of the address bus to define a full 16-bit address. The expansion memory is enabled for databus access when the processor addresses the overlay page.

This *porthole* or *window* expansion scheme can be further extended by using more page definition words; a second word expands the external address word to 24 bits and accesses 16 Mbytes of external memory. Access time is reduced by the need to load the two defining words before the page can be accessed. The scheme is best suited to accessing a large array of data which is accessed sequentially (see Chapter 9).

The Apple I/O architecture allows the scheme to be taken to dizzy limits. The I/O pages reserved for slots 1–7 can all be used independently as expansion windows. Addresses in the device select area of each slot can be used as expansion overlay buffers and pointers. Multiple banks of expansion memory can thus be accessed via individual peripheral devices. For historical reasons, this scheme is not used; the *language card* scheme described above is standard.

The early language cards for the Apple carried 16 Kbytes of RAM. To allow the bottom 4 Kbyte segment (hidden behind the I/O area, $C000–CFFF) to be accessed, a scheme was defined by which the 16K RAM chip could be mapped in one of two modes onto the 12K ROM space. This is shown in Fig. 8.17(a). This scheme expands easily to multiple arrays of 16K RAM chips (Fig. 8.17(b)) and, later, multiple banks of 64K chips. Whenever the expansion memory is active, the entire MONITOR and language ROM is disabled; any software that needs routines from there

Figure 8.17   (a) 16K RAM expansion, (b) 64K bank expansion.

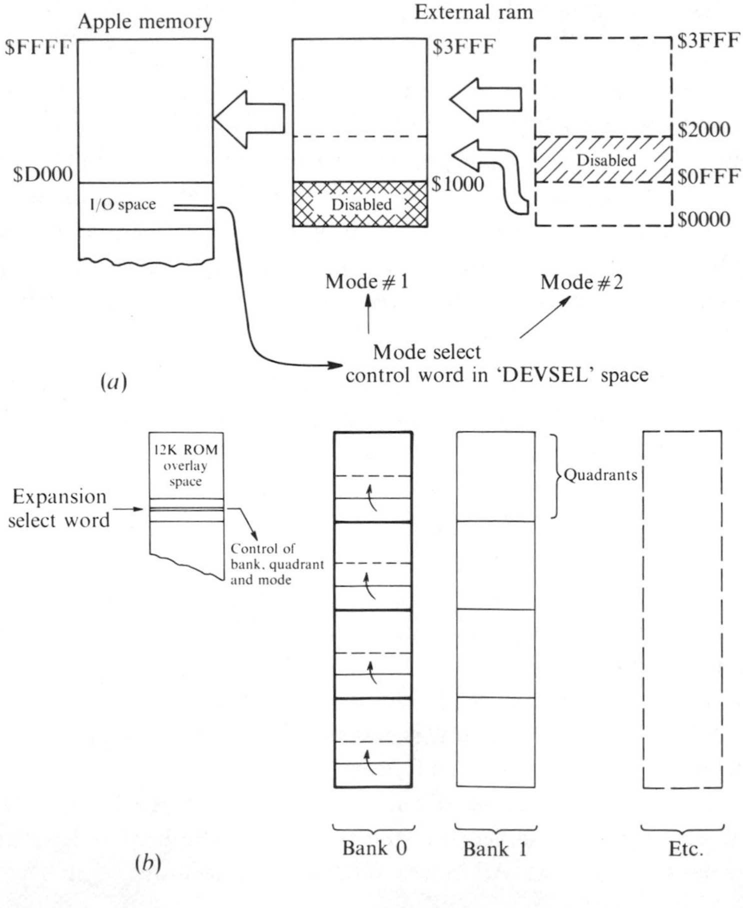

must either toggle the ROM back and forth, or must have a copy of the required routines in the RAM overlay. It is a bit messy.

No similar provision is made in the BBC machine for direct overlay of ROM space. The ROM decode logic is, however, connected through a patch-link block on the main board. The patch could be rewired to allow multiple 16K RAM banks to be overlayed in the ROM area.

A DOS implemented as described in Section 8.1.3 is really a form of page- or segment-oriented memory overlay. The DOS, on command, brings a page or segment of data from the disc medium into the buffer area in RAM. The user accesses it there; when the user is finished or when the next disc sector is needed, the DOS restores and/or overlays it, as required. A good DOS allows this to be carried out with a minimum of user fuss.

Extensive and elegant address expansion schemes are available for 16-bit microprocessors. They usually have powerful memory management support chips (Section 8.4.4). The 68000 processor provides a non-multiplexed 24-bit address bus for simple access to 16 Mbytes of memory. The Zilog Z8000 has a 23-bit (8 Mbyte) address bus. The top 7 bits are defined by a segment pointer within the chip. The physical address is defined by this pointer and a 16-bit offset address within the segment. Thus, the physical memory is arranged as 128 segments, each of up to 64 Kbytes. The 8086 only has a 16-bit address bus, but has four extra lines to provide a rather clever address expansion system. Four of the internal 16-bit address registers are defined to be segment registers for code, data, stack and a spare. The contents of these are used as an address pointer, but are shifted 4 bits left. They can therefore point to any 16-byte boundary in a 20-bit address map. All memory access by the 8086 is made relative to these segment registers, and all memory access is therefore dynamically relocatable simply by changing the segment register.

8.4.2 Operating systems

Contemporary computers conceal the intricacies of peripheral service routines behind simple system commands. The system user never needs, for example, to write code to collect keypresses, or to insert the bit map or text map into the display memory. These functions, and many more, are handled by resident software or firmware, the *machine operating system*, called-up by simple system commands.

The structure is built on several levels; Fig. 8.18(*a*) depicts these, and shows the levels of implementation in the Apple (Fig. 8.18(*b*)) and the BBC (Fig. 8.18(*c*)). In the Apple and BBC, the resident High-Level Language (HLL) is an *interpreter*, whereas the HLL in many machines is a *compiler*.

Figure 8.18   (*a*) Software and operating system levels, (*b*) the Apple configuration and (*c*) the BBC configuration. Machine language code can access all levels.

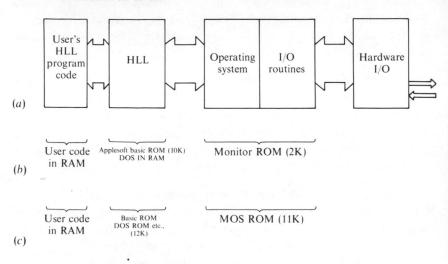

Although this book is not about HLL programming, the design of efficient interaction between the outside world and the computer for data collection and control requires an understanding of the differences between these two. Box 8.1 outlines the essential points.

Box 8.1
Interpreters and compilers

There are two classes of high-level programming languages. Both require the user to observe a clearly defined protocol for the definition of variables, operations and commands. There is a vocabulary of commands and operations, and a strict syntax for their usage and combination into HLL program code.

An interpreter language such as Microsoft BASIC collects the user's HLL program code and stores it almost exactly as entered from the keyboard, tape or disc. If the code is entered from the keyboard, the keypress string is assembled in the keyboard input buffer until the 'carriage return' key is encountered. The resident language code will then scan the buffer. If the first character is a number, the string is deemed to be a line of program for storage in the user program memory. Language commands (such as PRINT, IF, THEN, LET etc.) are then coded as single bytes to save storage space, and the coded line is stored.

If the first character in the buffer is an alphabetic character, the HLL program codes the string and passes it directly to the command string interpreter routines. When the user's command and operand string is to be executed in this IMMEDIATE mode, or when a program in memory is run, the code is scanned by the HLL INTERPRETER to break it up into HLL COMMAND strings; these are decoded and machine language subroutines are invoked to carry out the commands. An interpreter program is thus able to execute the user's code in IMMEDIATE or DEFERRED (program) modes.

The user's program code is held in memory in an easily accessed form. This makes it very convenient for writing and debugging user programs; changes can be tested as soon as written. However, the execution of the code requires that it be fetched, scanned and converted to machine language routines every time it is run. A lot of processor time is required for this lengthy process.

Interpreter BASIC is very convenient for working up interfaces and peripherals, since it has simple direct commands to access memory/I/O addresses, and it can execute IMMEDIATE commands. This allows the interface hardware to be tested from the computer keyboard.

An interpreter program (such as Microsoft BASIC) is the resident HLL in almost every personal microcomputer that does not require a disc-based system. Microsoft developed the compact interpreter in the late 70s, for the earliest microcomputers. It was based on the simple, easily learnt structure of the multi-user time-shared BASIC language in universal use at that time, for undergraduate programming tuition on mainframe computers. With considerable ingenuity, and by making big compromises in execution speed, they managed to fit the core of the language into a few kilobytes of machine language code. Such a small program, with a versatile command structure and a lot of friendly features and error messages, could only be implemented as an interpreter.

The other form of HLL is called a *compiler*. The user's code is written under the HLL protocol, but is not used directly in this form. The code is treated as text and must be written with some form of Text Editor. If the HLL is line-oriented (e.g. BASIC), the editor must be line-oriented, so that lines can be fetched, modified, moved around etc.

When the programmer is satisfied with the code, it must be passed through a compiler, which scans it in much the same manner as an interpreter. It does not, however, execute the user's commands. Instead, it converts the user's source code into a machine language program. The user's variables, operations and commands are translated and compiled into a new program which accesses variables directly in memory and specifies jumps to internal subroutines or introduces blocks of machine language code to carry out the operations. The result is the final, *compiled* machine language program, usually called the object code. This code is self-contained and can run on its own.

It is very difficult to break into the compiled program to make small changes. If any change is needed, the source code must be reedited and the compilation must be repeated. Program workup and debugging is therefore very tedious.

Compilers tend to be rather big and complicated programs. (Since the compiler need not be present when the compiled program is run, there is little incentive to minimise its code.) Compilers thus tend to fill up available memory in the machine with fancy features. They are usually run under DOSs; the user's source code is fetched from disc as a file and the compiled code is returned to disc as another file.

The obvious advantages of both forms of HLL can be merged if code-compatible interpreter and compiler languages are available. The workup and debugging of programs and interfaces can then be carried out on the friendly but slow interpreter, but the final version of the source code can be used to generate a fast, compiled object code.

In many research applications, there are many standard complicated procedures (such as a Fast Fourier Transform or a polynomial transformation of an array) that need to be carried out on collected data. Data analysis frequently involves different programs or sequences of applications of a suite of such procedures. In these situations a very powerful compromise between interpreter code and compiled code can be applied. The *control* program, defining the sequence of operations, is written in the interpreter HLL but the complicated number-crunching of the procedures is carried out by fast, compiled routines. This scheme requires that the interpreter is able to pass variables and control parameters to the compiled routines; thus it requires that complete, clear, descriptive documentation is available for both the compiler and interpreter HLL. The Microsoft Interpreter BASIC and Compiled FORTRAN operating under MSDOS can be used thus without serious problems. The documentation is a bit obscure, but has all the necessary information. It also forms a very convenient system for merging interface service routines and powerful processing of the data collected.

CPM and MSDOS are disc-based operating systems written for a specific processor (or processor family). Operating systems such as these are made essentially machine independent by being divided into two distinct parts. The major part handles all the high-level system functions, while the minor part interfaces the software operating system to the specific I/O hardware interfaces of the host machine. This minor part is usually called the *Basic I/O Service* (*BIOS*) module. The operating system is adapted to any computer by modifying the BIOS interface routines. (Compare this to the processor independent code concept described in Section 5.4.) The 'shell' structure is depicted in Fig. 8.19.

User programs can access the structure at any level. It is easiest from the HLL level, but user machine language code can use the BIOS routines for indirect, machine independent, hardware access. This lovely feature is even nicer if the interface service routines are arranged within a software interrupt structure (see Section 7.6.3). The programmer need then only know the interrupt vector and the procedure for passing parameters.

8.4.3 System bootstraps

The simple bootstrap procedure outlined in Sections 6.6 and 7.6 needs to be elaborated somewhat to handle the boot-up of complicated systems.

The Apple II Plus will boot directly into the resident HLL unless a disc interface is present; in that case it will attempt to introduce the DOS system. A section of the ROM boot scans the interface slot I/O pages for the code in the disc ROM. If found, the processor is passed to this program; it is a short routine to force the disc head to Track #0 and to load Sector #0 into RAM and transfer the processor to this code. This process continues on to load the entire DOS code and link it to the MONITOR I/O routines

Figure 8.19   The shell structure of CPM, MSDOS etc. and user programs.

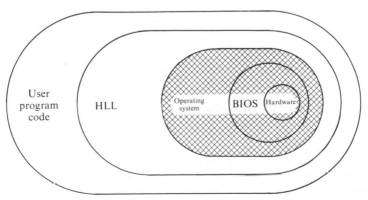

so that DOS commands can be intercepted in the input or output text stream.

The final stage of a DOS boot-up loads and runs the first user program in the directory of the boot-up disc. The Apple DOS EXEC(ute) command causes a text file on disc to be read as if it were a sequence of IMMEDIATE keyboard commands; thus any procedure or sequence of DOS commands can be executed automatically at boot-up.

Boot-up of a Z80 system is a bit different. The RESET control line of the Z80 clears the PC (and other internal registers) to $0000. When the RESET is lifted, the Z80 attempts to fetch an instruction from address $0000. If the machine has a full 64K of RAM memory, some hardware trick must be used to steer the processor into the boot-up ROM.

Sophisticated interrupt structures are invariably provided by 16-bit microcomputers; RESET is usually handled, like an interrupt, by fetching a power-up vector. The reader is referred to the handbooks and data manuals for details.

Resident system firmware in ROM makes boot-up fairly simple, but pins down the operating system unless ROM bank switching is provided. Disc-based operating systems such as CPM and MSDOS have a more versatile structure. The entire operating system is brought into RAM from the system disc. All that is needed in the host computer is a small ROM segment in the disc drive hardware to get the system bootstrap loader into RAM and to point the processor at it. The bootstrap then brings the entire operating system into RAM and initialises it. HLLs such as BASIC, FORTRAN, PASCAL etc. can then be loaded above the operating system. The user can bring these from disc into his own machine as part of his system structure, and can then use them to write programs in the newly resident HLL. The resident operating system allows machine independent I/O operations by passing all input and output through the machine specific BIOS routines; this machine-independent structure of the operating system is intended to ensure software compatibility between different machines.

The IBM PC has the BIOS in ROM, on the main processor board.

8.4.4 Memory management

Mainframe and large minicomputer systems are usually set up in a multi-user or multi-task configuration. The machine can be accessed from several terminals; each user generally requires very different software and firmware structures. The operating system for this configuration must be quite complex, since it must handle all of this extra organisation.

A simple technique used in early machines was to partition the available memory and to set up the different operating systems needed by each user within their own partitions. The memory is more efficiently used, however, if one resident operating system is accessed by every user. A large variety of software packages for common tasks can then be shared by all users.

To simplify programming, every user must be able to access (link into) the operating system and the resident software packages. In a single-user situation (as in personal microcomputers), the linking is quite straightforward. The user simply nominates the packages required and the operating system loads them into memory and sets up the linkages.

In the multi-user situation, the entire space appears to be available to each user. This is handled by a *memory management system* which installs in the actual machine only that part of the user's program which is being used; the rest is kept outside on disc.

The virtual memory structure, the constant *logical memory* arrangement seen by the user, is very different to the actual, continually changing *physical memory* usage. Since several users or programs may be simultaneously using the same area of the logical space, the memory management system must be able to relocate dynamically the user's code to different physical areas of memory. This very powerful technique can be extended to include the back-up memory (hard or floppy disc) as part of the virtual memory seen by the user.

If the operating system can anticipate the needs of the program, it can transfer data and code into memory while the processor is still executing the preceding code. This is called pipelining since the code and data is coming 'down the pipe', to be in place before it is needed. The operating system overhead can only be minimised if the memory management is done as a background task and seldom gets in the way of user programs. The importance of block movement under DMA control is obvious.

Multi-tasking provides for several different programs to be resident (virtually) within the machine at one time. Program execution can be passed from one to the other, and a common data space can be set up for parameter and data passing.

Sophisticated multi-user or multi-task operating systems continually monitor the usage of memory segments; when a segment is being little used, it is moved back to disc and is replaced by a more important segment. This is intended to yield maximum efficiency of memory use; if several programs all have very variable patterns of memory access, the system can waste a lot of its time trying to optimise the rapidly changing usage.

The 68000, Z8000 and 8086/8 are all supported by Memory Management Unit (MMU) companion chips. These provide very powerful virtual memory expansion. Manufacturers' data manuals provide extensive information and application notes.

8.4.5   Mass storage, back-up memories and errors

The virtual memory concept allows indefinite expansion. A large virtual memory array heightens the probability of errors due to either random noise or simple bit failure. Any error in the processor's instruction code is fatal and a single bit error in data can have disastrous consequences to a computation. Large memory systems must therefore be provided with some form of error detection system. If the system must keep running when an error is encountered, some form of error correction must also be provided.

An early solution to the error problem, used in the first automatic instrument landing systems for commercial aircraft, was the *Triplex* system with three independent control systems. The outputs of all three were continuously *polled*, the majority opinion was considered correct. While all three sets of outputs agreed, it was extremely unlikely that any were in error. If one output disagreed, that system could be deemed faulty. Safe operation and fault detection were thus almost completely assured.

This solution is too expensive for general computer use. Simpler solutions have been devised for the general problem of error detection and correction within the realm of digital electronics. The simplest form of error detection uses the concept of parity, as discussed previously in Section 2.4. It is used in the form of the checksum (as described in Section 8.1.3) to provide a simple, quick check for the integrity of blocks of information.

Error detection can only be achieved by the presence of some redundant information. Error detection *and correction* can only be achieved by providing extra redundancy. The Triplex system is an obvious, extreme case.

The work of Hamming in the 1940s, and subsequent developments, have allowed the reliability of redundant error-prone systems to be defined precisely. *Hamming codes* with different levels of redundancy have been defined for optimally efficient error detection and correction to any level of reliability. The error detection and correction codes are available, of course within microchips! The Intel 8206 error detection and correction chip, for example, handles a 16-bit data word stream (with up to 8 additional check bits) and can be cascaded in parallel up to 80 bits. It uses a modified Hamming code to detect and correct all single-bit data or check errors. It can detect all double-bit errors and most higher multi-bit errors. This type of error correction chip is now regularly used for very large DRAM memory banks. A 1 Mbyte memory board is now quite economical with 64K chips. A typical board organised with a 16-bit data word would use 128 chips for data and 48 more for check bits (6 per 16-bit word). Organised as a 32-bit word memory, only 28 check chips would be needed (7 per 32-bit word).

The ubiquitous floppy disc has a moderate capacity and, since it is fairly fragile, it cannot be spun very fast. The maximum capacity for an 8″ double-sided disc is about 1 Mbyte, and the worst-case access time to a particular track and sector can be hundreds of milliseconds.

Much better capacity and response time can be achieved with a hard-disc system. The general principle of operation is much the same as the floppy disc, but the disc is spun at a much higher speed. The R/W head is kept a few microns clear of the rigid media surface to eliminate rubbing and wear. Access within a track is very fast and the media is virtually everlasting unless the head crash-lands on the disc. (Back-up on magnetic tape is mandatory.)

Tracks can be spaced very close together on the precisely centred rigid disc. Typical track spacing is about 5 μ compared with the 250 μ needed on a simple floppy system. A typical capacity for a small single hard-disc system is 10 Mbytes. The very fast data transfer rate possible with the very high disc speed makes the hard-disc system ideal for virtual memory applications with DMA data transfer between disc and memory.

Magnetic bubble memory systems have been available for several years. They were hailed as the memory of the future, but they have failed to fulfil this expectation; production problems seem to be the reason. Their principal appeal is that they have no moving mechanical parts and are non-volatile (like any magnetic storage media, they retain their data when they are turned off).

The storage medium is a very thin layer of oriented, magnetic material with its allowed magnetic axis normal to the film. Under a suitable magnetic bias field, the spontaneous magnetisation in this material opposed to the field separates into small domains a few microns in diameter. These small domains or *bubbles* will move up a magnetic field gradient. Hence they will attach themselves to a ferromagnetic overlay laid down above the thin film (see Fig. 8.20(*a*)). If these overlays are of a suitable shape and are replicated around a track, a small rotating magnetic field applied to the whole system will make the bubbles circulate around the tracks. If a pattern of bubbles is placed in the track, it will circulate continuously while the driving field rotates, and will be stored when the field is stationary. The memory system contains many of these loop tracks arranged between an input and an output track (see Fig. 8.20(*b*)). A *seed* bubble, held circulating under a small permanent disc magnet at the start of the input track can be controlled such that it will generate a binary sequence of bubbles (i.e. bubble or no bubble) along the input track. Swap blocks at the input end of each loop can be activated by a timed current pulse in a loop beneath the blocks to divert the input track pattern into the loop and to divert the old loop data out into the remainder of the input track to be annihilated at the end. Replication blocks at the output end of each loop can be activated by current pulses to split the circulating bubble into two parts (the circulating bubble regenerates to its original size). The loop replica generated by the replication block passes along to the end of the output track where it is detected and amplified to become the

Figure 8.20   Magnetic bubble memory: (*a*) the arrangement of the ferromagnetic overlays on the thin film, and (*b*) the system of loops and tracks needed to load, store and read data.

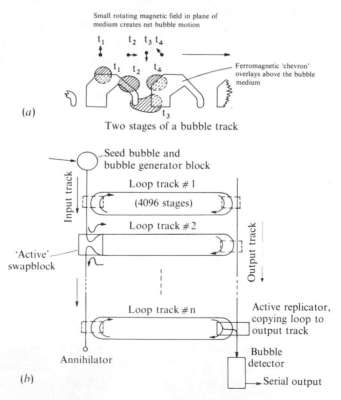

digital output stream. The circulating field runs at about 50 KHz; with 4096 bits per loop, the access time can be up to 80 ms.

Unlike the magnetic disc, where the magnetic media holds stationary data but moves under the R/W head, the bubble memory moves the data pattern over a static medium from the write to the read heads. A typical device contains 256 tracks each with 4096 bits each. The nominal capacity is 1 Mbit (128 Kbytes), about the same as a floppy disc. An equivalent capacity (volatile) DRAM system would need sixteen 64K, four 256K chips or one 1M chip. The entire system, with its complex drive electronics and interface buffers, is about 10 cm square and about 1 cm deep.

Video cassette recorders have the potential to hold enormous amounts of binary data, but the quality of the media is not suitable for high-integrity data unless a lot of data redundancy is added. They are used for back-up of hard discs, or for very large serial-access databases. Access time to random data is very long.

The most promising new medium for large-scale data storage undoubtedly is either the compact digital hi-fi disc or the digital video disc. Both use a finely focussed laser beam to scan a digital data track consisting of small pits or gaps which modulate the reflected laser light intensity; the serial digital information collected from the reflected laser light is used to regenerate high-quality sound signals or a video picture. A servo control system keeps the laser beam tracking accurately along the data path.

The discs were originally intended only for players; reproductions of a master can be made by a refinement of the conventional analog disc moulding process. The Japanese 12 cm-diameter compact hi-fi audio disc is now available in several modified players as an Optical ROM (OROM) with a capacity of several hundred megabytes. Several American manufacturers are offering similar capacity of OROM on a 5.25"-diameter disc. IBM is expected to offer a small add-on peripheral 2"-diameter OROM system with a 40 Mbyte capacity for its PC machine. This is intended to simplify the loading and running of its version of the UNIX operating system, PC-IX.

Several Japanese manufacturers have demonstrated optical disc systems which allow data to be written into the optical track. Optical disc blanks with special surfaces can be written by a modulated laser beam which either ablates pits in the disc surface, or generates blisters in a heat-sensitive layer just below the disc's reflective surface. Both of these systems yield *write-once/read* memory. This is no problem for archival data storage since the enormous capacity of the disc allows it to be used for normal editing and storage work. The operating system can delete a file by removing the segments assigned to it from the file directory and by locking out that segment from further use.

Subtler storage techniques are also being developed for R/W storage. In one of these the data is stored by a reversible crystalline–amorphous transformation of the recording surface, with a concomitant change from reflective to transparent. Philips and Matsushita are developing magneto-optic systems where the information is stored magnetically, but is read and written by a polarised laser spot. The polarisation of the reflected beam is affected by the magnetisation of the medium.

The compact audio disc uses simple solid state diode lasers focussed to about $1\,\mu$. The data is recorded on a spiral track of about $2\,\mu$ radial pitch. A digital video disc holding one hour of colour video, contains several gigabytes of information; the disc is therefore able to store gigabytes of data. The data density requires a submicron spot focus and, so far, can only be achieved with gas lasers. Eventually, they should provide an order of magnitude capacity improvement over the systems now evolving from the compact audio disc.

The video discs, like hard magnetic discs, spin at quite high speeds (1500 rpm for the 50 frame European standard) to scan one full frame per turn. The freeze-frame feature now in vogue for video playback requires that the read head is able to lock and track on one spiral turn. If the head control mechanism were reworked to provide individual annular tracks, a superlative mass storage system with very fast data access would be available.

8.5   The first 40 years, and the future

Before the invention of the transistor, all electronic operations were carried out by thermionic valves. These devices were massive, expensive and required a lot of power to keep their cathodes hot (about 2 W per valve), hence the machines had to be designed to minimise the total number of devices. Unlike the multi-bit logic processors discussed throughout this book, the first-generation machines processed data in a bit-serial stream passed through a single gate network.

The basic logic elements were simple binary electronic pulse circuits which performed as flip-flops or as *threshold logic gates*. A multiple-input threshold gate outputs a pulse only when a predefined number of its inputs are pulsed simultaneously. A multiple-input gate with a threshold of one represents the **OR** logic function; an N-input gate with a threshold of N represents the **AND** function. *Inhibit inputs* are equivalent to negation. The layout and operation of this type of logic is very like that of the synapses in our neural system. (See the brief remarks at the end of this section.)

Combinations of the basic gates can generate all logic functions. The logic diagram of a serial full adder using threshold logic is depicted in Fig. 8.21(a). The addition of a delay element to store the carry output for one bit period of the serial word, allows one such circuit to carry out full addition

Figure 8.21   Serial-bit stream: (a) addition; (b) multi-bit full addition and (c) multiplication.

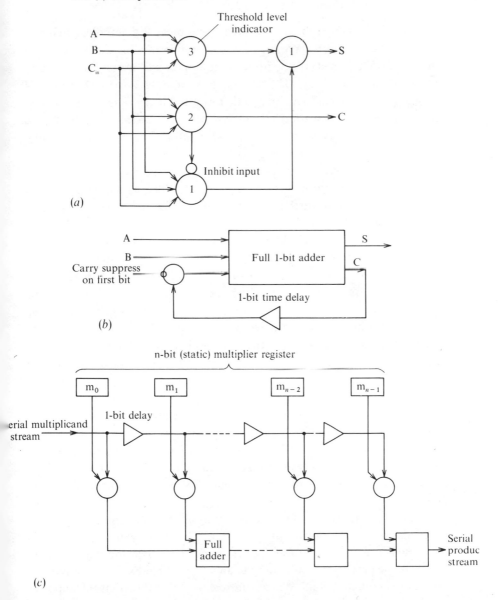

to any word size (Fig. 8.21(*b*)). The two arguments are fed in l.s.b. first, the serial sum streams out simultaneously with the input.

Multiplication is performed quite efficiently but requires a full-word flip-flop register holding a *static* representation of one argument (Fig. 8.21(*c*)).

Some form of fast *serial* storage was needed to hold transient data. Acoustic delay lines using mercury columns or taut nickel wires were commonly used. Serial data streams could be loaded and circulated indefinitely or fetched when needed.

The invention of the transistor allowed economical, fast, parallel machines. Serial-access memory gave way to fast, random-access, parallel word memory. Bulk data storage within the computer was usually provided by magnetic core memory. This stored bits of data as the direction of magnetisation in small ferromagnetic toroids. The toroids were moulded from magnetically hard ferrite material with a wide magnetic hysteresis loop; the toroid was readily magnetised in either direction by a nominal current I_M as shown in Fig. 8.22(*a*).

The toroids were arranged in columns of words. A simple R/W arrangement is shown in Figs. 8.22(*b*) and (*c*). The actual configuration used was more elaborate than is shown, but the figure is adequate to show the principles.

Each row (word) was linked by a common word line and each column (data bit) by a common bit line. The array was read by pulsing a current of I_M through the selected address word line. If a particular toroid in that word was previously magnetised in the opposing direction, its magnetisation would be reversed and a voltage pulse was generated in the bit line. If no flux reversal occurred, the pulse in the bit line was due only to the mutual inductance between the wires, and was very small. Thus the flux pattern present in the row could be detected and latched for output as a logic word.

After the read pulse, all toroids in the row were magnetised in the same direction (a *destructive read*). The data had to be rewritten immediately (or modified data entered). This was done by using the column lines to drive a reverse current of $I_M/2$ through each bit line which was to be rewritten as a logical **1**, and to then pulse a reverse current of $I_M/2$ through the word line; only those toroids where *both* currents flowed would have their flux reversed.

Once written, the magnetisation pattern persisted indefinitely and needed no further power or attention. This *non-volatile* feature was exploited in many computers by arranging that when the mains supply went off the processor would jump to an interrupt *power-down* routine that stored the contents of all registers in reserved core locations. (The power supply had enough stored energy to run the machine for a few milliseconds.) When power was restored, the memory was exactly as it was before power-down; the power-up interrupt routine reloaded the processor registers and the computer resumed operation at the interrupted instruction.

The arrays were arranged as *memory planes* of typically 1K × 16 bits. The read/modify/write cycle required about 1 μs. The toroids were about 2 mm inside diameter with a cross-section of about 1 mm square. The lacing was done by hand, usually in developing countries where labour was very cheap.

Pneumatic logic devices were introduced by Schrader and Dow-Corning in

Figure 8.22   (*a*) Hard ferromagnetic magnetisation curve, (*b*) a single toroid and (*c*) toroids assembled into a memory plane.

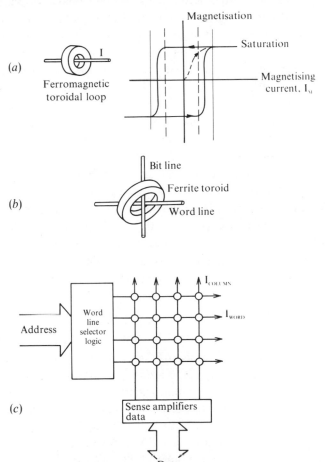

the 1960s. Simple boolean operations and flip-flops were available. They were used for control systems in hazardous environments such as mines, where electromechanical systems were dangerous.

The Schrader system used a thin laminar-flow jet of gas which passed through the output unless an 'input' jet perturbed it. This was a **NOR** gate, with a considerable transition delay. The Dow-Corning system used the Coanda wall-attachment effect between a thin jet and the two walls of a triangular duct. **NAND**, **NOR** or flip-flop gates were available. Dow-Corning built a simple pneumatic computer from these. It worked, but its instruction execution time was about 20 ms.

The entire development of electronic computing has taken place in less than the working lifetime of one generation; the exploitation of transistors in computers has been within a lesser time; the computer revolution, driven by the rapid development of microcircuits, has taken place within only 10 or so years. There has not been enough time for the science and technology of computers to be reviewed thoroughly and to reach maturity. The apparent sophistication of the present machines is due to their complexity, rather than their subtlety. The basic architecture of computers has not developed far beyond the original concepts of Babbage, Turing and von Neumann. Contemporary machines, based on these old concepts, have become very fast, complex and powerful; they can handle many applications superbly, but there are many which they handle very badly or cannot handle at all.

In spite of the enormous processing power available in the latest machines, it is still extremely difficult to make a computer carry out many of the tasks handled effortlessly by the human brain. A young child has an innate ability to discern the complex and subtle protocols of language, syntax and vocabulary. We humans also have a remarkable ability to analyse sound and light patterns, to collate complicated experiences, to classify and store these in cross-referenced categories, and to use our accumulated database at a later time in novel combinations, for intelligent or apparently original thought.

Current generation computers are only just adequate, with very extensive and complex software, to create an *artificial intelligence*, where the software gives the impression of having some human intelligence. Expert systems formalise the knowledge of a human expert in some field into a well-organised, linked, information tree structure. A non-expert can then be led through the information structure to find (already-known) solutions to problems. The system has no innate ability to devise new solutions, and is still a long way from the true meaning of intelligence.

Artificial intelligence systems still have a long way to go before they can be said to have *synthetic intelligence*. There is, in fact, some doubt that such intelligence can be created with existing computer architecture and technology.

Current research and development on computers is aimed at either providing more computing power and speed, or at solving the enormous and subtle problem of synthetic intelligence. One path leads to further development of the existing architecture to achieve ever-faster program execution. More powerful machines are needed to handle the number-crunching and data processing used to model systems with many interactive variables. These include climatic models, economic models and, unfortunately, global weapon systems. The problems here are predominantly technological and are certain to be solved. Let us hope that we survive their ultimate solution. Ultra-fast logic and memory is already used in machines such as the Cray supercomputer; the Cray 1 has a cycle time of 12.5 ns, the newer Cray X-MP has twin processors that cycle in 9.5 ns. The Cray 2, due for release in 1986, will have four processors working at twice the speed. The Cray 3, bigger and faster, is under development.

A lot of research effort is also going into optical logic devices. These have a potential for switching times of picoseconds; if they can be realised, instruction execution speed will become an insignificant limitation on computer operation.

Enhanced computation speed can also be achieved in a suitable architecture by making several conventional processors work in parallel. An obvious architecture is to make them all share access to a common memory holding programs and data, but with each processor working simultaneously on different tasks in the overall program.

Dataflow machines use a technique where the algorithm for a large task is broken down into several paths which can be executed simultaneously; the evolving data flows through several processors and comes together at the end of the task. Programming requires specialised programming languages; the problems are not yet fully solved.

Another technique, called *Distributed Array Processing* (*DAP*), developed by GEC, uses a large array of identical processors which all work with the same instruction sequence, but operate on different data elements in an array. This architecture has considerable potential for applications such as optical image analysis; it can model, at a very simple level, some of the preliminary data processing that is carried out by the neural network in the retina of insect and vertebrate eyes.

INMOS have developed a *transputer* system where the individual single-chip computing elements, the transputers, act as parallel processing elements in the system or can operate independently on a common database. All of the transputers communicate with each other over a simple, high-speed, serial data path; their instruction set includes synchronised communication commands. This eliminates the bus problems that arise when arrays of conventional processors are connected via conventional bus systems. The hardware has been developed in conjunction with a programming language OCCAM which exploits all of the architecture's capabilities.

Much of the development work in Japan on the coming *fifth generation* computers, is attempting to cram enough computing power into known microcomputer architectures to allow speech patterns, common language vocabulary, grammar and syntax to be analysed in real-time. The intention is to provide a spoken language, conversational interface between the machine and its operator. Automatic translation between written or spoken languages will evolve from this, Workers in these fields seem to be confident of a solution within existing or anticipated architectures and technology. The necessary power seems to be achievable with parallel processing and data flow architectures.

The alternate paths of development are much more difficult. They require the invention of new architectures which allow a new approach to synthetic intelligence. The principal problem is that intelligence is not yet properly understood.

One obvious difference between a computer and a living system is the enormous amount of parallel processing that is carried out at all synaptic levels of the nervous system. From the primary sensors through to the very high-level perception processes in the conscious brain, every level seems to be involved in temporal or spatial data processing or pattern recognition. The output from each successive level corresponds to increasingly sophisticated processing of the original sensory input. The retina of the eye processes the visual field data from the primary mapped-image into an elaborate but highly processed set of specific coded signals indicating the motion and orientation of contrast contours in the image field. Higher-level synaptic processes in the brain process the complex data stream into recognition of a friend's face, a sign of danger or whatever. Multi-level parallel processing seems to be involved in all sensory input systems; it seems reasonable to conjecture that it is also the basis of conscious thought and of intellectual reasoning. The architecture allows the very slow basic nerve process, the action potential wave, to achieve rapid, sophisticated control of the animal body and the intellectual mind. A conventional computer with a logic-signal transition time of a millisecond or so, would not be able to perform so well.

Computers cannot yet achieve anything like this level of parallel operation. Some tentative first steps have, however, been made. The DAP, mentioned briefly above, models the simplest level of retinal processing. Semionics (USA) have developed *associative memory* systems which locate data in memory by content rather than location. ICL (UK) has developed a *Contents Addressable File Store* (CAFS) which provides simple processing ability within a disc controller system; the controller is able to sift quickly through its files looking for patterns without the need for main processor involvement.

Aristotle's logic formalism was introduced in Section 1.2, but was only followed to the level of binary combinational logic. Conventional computer architecture is based entirely at this level, and on its extension to sequential logic. The traditional formal logic, however, provides a higher level of sophistication. Its purpose was to formalise the logic processes involved in the *inferences* inherent in common language, and the processes of logical thought and reasoning. It does not attempt to analyse 'original thought', but it allows the truth value of compound statements or propositions of common language to be inferred from the truth of the simpler propositions on which they are based. This formalism should thus be able to model intelligence more efficiently than the simple combinational logic on which current computers are based. A lot of effort is being put into developing *inference machines* which operate at the propositional inference level. At present, the performance of possible architectures is being tested by modelling them in conventional fourth-generation machines.

The present indications are that, to provide useful synthetic intelligence, these new machines will have to be capable of millions of inferences per second; the emulation software can only manage about a thousand per second.

Programmable computers have evolved in less than 100 years, whereas the vertebrate brain has had quite a lot longer to make its architecture efficient. While the evolutionary pressures have been very different, better computer architectures may have to be modelled on some of the more powerful and, let us hope, benign, features of the human brain. Provided, of course, that we have the wit to unravel them.

References and further reading

1 *The Apple II Circuit Description*, W. D. Gayler, W. H. Sams, Indianapolis (1983)
2 *Principles of Colour Television Systems*, C. R. G. Reed, Pitman, London (1969)
3 *PAL Colour Television*, B. Townsend, Cambridge University Press, Cambridge (1970)
4 *NEC APC Reference Manual*, NEC Information Services (USA) (1983)
5 *EF9367 Graphic Display Processor Application Principles*, G. Lambert, Application Note NA-036A, Thomson-EFCIS (France) (1982)
6 *Beneath AppleDOS*, D. Worth & P. Lechner, Quality Software, California (1981)
7 *Computer Communications*, R. Cole, Macmillan, London (1984)
8 *The Electric Telegraph – an Historical Anthology* (ibid. Chapter 1)
9 *RS-232 Made Easy*, M. D. Seyer, Prentice-Hall, Englewood Cliffs, New Jersey (1984)
10 Standards:
   RS-232/-422/-423/-449, Electronic Industries Standards Association, Washington (USA)
   IEEE Std 488, Institute of Electronic and Electrical Engineers (USA)
   CCITT V24 Standard, International Telecommunications Union, United Nations (USA)
11 'The IEEE Standard for the S100 Bus', M. Garetz, *Byte*, Feb. 1983, pp. 272–98
12 'Welcome to the Standards Jungle', I. H. Witten, *Byte*, Feb. 1983, pp. 146–81
13 'The Hewlett-Packard Interface Loop', R. Katz, *Byte*, Apr. 1982, pp. 76–92
14 'HPIL', R. D. Quick & S. L. Harper, *Hewlett-Packard Journal*, Jan. 1983, pp. 3–7
   References 1–14 cover Section 8.1.
15 Data manuals for microprocessors (available from their manufacturers)
16 'Design Philosophy Behind Motorola's MC86000', Pts 1 and 2, T. Starnes, *Byte*, Apr. 1983, pp. 70–95; May 1983, pp. 342–67
   References 15 and 16 cover Section 8.2.
17 Memory data handbooks from Intel, National, Fairchild, Motorola, Siemens etc.
18 *National Programmable Logic Databook* or the *Fairchild TTL Data Book* (Vol. 2)
   References 17 and 18 cover Section 8.3. The databooks in Reference 18 cover PAL and PLA arrays.
19 'Virtual Memory for Microcomputers', S. Schmitt, *Byte*, Apr. 1983, pp. 210–41
20 *Intel Memory Components Handbook*, Intel (USA) (1983), Application Notes AP-46/73, AR189/197

21 'The Reliability of Computer Memories', R. J. McEliece, Scientific American, Jan. 1985, pp. 68–73
   References 19–21 cover Section 8.4.
22 'Artificial Intelligence', Theme for *Byte*, Apr. 1985
23 'Robotics', Theme for *Byte*, Jan. 1986
24 *The Brains of Men and Machines*, E. W. Kent, McGraw-Hill (1981)
25 *Byte, New Scientist, Scientific American* and other such journals have occasional news and review articles on computer developments
   References 22–5 cover Section 8.5.

9

Instrumentation and data-collection applications

Preamble

Chapter 7 covered the basics of interfacing and interaction between a microcomputer and the outside world for conventional data-collection and control applications. This chapter describes some examples that demonstrate how simple interface hardware turns a microcomputer into a powerful data-collection or control system. Limitations imposed on a real-time system by the architecture of the computer are explained, and further examples show how these limitations can be overcome. The examples are, without apology, taken from systems developed by the author. Most of them involve applications in Physics research, where microcomputers have had an enormous impact.

The rapid growth of Science and Technology in the last hundred years has taken research a long way from the 'pen and notebook' age. Our continual desire to improve our conceptual models and understanding of the machinery of Nature has pushed research towards ever more sophisticated instrumentation, designed to sift small 'needles' of information from the 'haystack' of irrelevant data or noise.

The raw information is usually a set of readings of one or more observables of the system under study, as a function of time, space or state. Sometimes it must be gathered and recorded quickly for subsequent processing and analysis 'off-line', as in the Voyager and Giotto missions, or, alternatively, it can be processed in real-time as it is collected. (This is essential, of course, for a real-time control system.)

Research, like all other human endeavours, must share society's limited resources; it must always make the best possible use of available funds. When a field of research is developing quickly, an expensive new instrument can quickly become obsolete, but, worse than this, its presence may restrict the direction of subsequent research to suit its capabilities.

Since their inception, computers have been used in research for data processing and analysis but, if the computer is regarded as a programmable data-collection instrument rather than just as a black-box data analyser, it can become a major instrument in the research program. If a research team has an adequate knowledge of microcomputer hardware to design and build new interface equipment, a microcomputer can be reconfigured rapidly to follow the research program as it evolves. The program is then free to roam wherever the results take it, and at minimal cost. Much the same conditions apply in technology and engineering; process or product development can be best kept up to date by a team with a thorough working knowledge of microcomputers.

9.1   Instrumentation equipment and techniques

This section discusses standard, 'off-the shelf' modules and systems, and the alternative of building special-purpose interfaces, for microcomputer applications.

Many sophisticated electronic instruments are available with interfaces to one of the standard communication systems such as RS-232 or IEEE-488. This eliminates the problem of direct connection to the multiplicity of bus structures used in mini- and microcomputers, but requires that the computer also has an interface to the same system, and has the software needed to support it. Communication standards allow equipment from different manufacturers to exchange data, but require extra logic hardware and extra steps in the data transaction sequence to observe the communication protocol.

RS-232 requires dedicated logic at both ends of its link to convert between parallel and serial data and to handle and control the transaction protocol. RS-232 and its faster successors have been popular with software users because their serial transmission can be linked with relative ease to the telephone system.

The IEEE-488 bus system discussed in Section 8.1.4 is rapidly becoming the standard for local data collection and control in Science and Engineering. Most of the newer microcomputers aimed at the scientific and engineering market have an IEEE-488 interface fitted as standard or as an option (see Section 8.1.4). A wide range of instruments is now available with an IEEE-488 interface.

Standard communication systems allow quick assembly and commissioning of a data-collection or control system, provided that support software is available and that it can be linked into the HLL used in the computer. Improved performance can usually be achieved, however, by a data-collection system integrated directly onto the bus system of the host computer. The following subsections discuss this.

9.1.1   Computer bus systems

During the rapid growth of the microcomputer industry, there was a strong tendency for the manufacturers of microcomputers and peripherals to establish *de facto* standards for communication at the microcomputer bus level. The S100 bus is an early example. It was developed originally for a microcomputer kit using the 8080. It was not tightly specified in its original form, but it has since been tightened up as the IEEE-696 standard. Subsequent enhancement has included address and data expansion to accommodate 16-bit microprocessors. A wide range of peripheral equipment is available; it is aimed principally at the hobbyist field.

The STD bus system, devised and developed by Prolog Inc., is intended as a universal system to link specialised functional modules for data collection and process control. The bus protocol is tightly defined, with the result that all approved equipment is truly compatible. Most of the 8-bit microprocessors are available as plug-in cards with a resident operating system. Expansion memory, disc systems, display systems, development systems, PROM burners and a good range of instrumentation and control modules are all available at quite moderate cost. Straightforward data-collection and engineering process control applications can usually be assembled immediately from standard modules and programmed in a wide range of languages on any suitable microprocessor. It is also a very convenient framework within which to build special-purpose modules. The finished modules will work with any STD system and can be marketed through Prolog.

The E78 Europa Bus uses the standard Double Eurocard; the bus assignments are flexible and can work with 16-bit processors. Data transactions are carried out with full asynchronous handshake protocol at LSTTL levels on a standard 64-way connector. A second 64-way connector allows user defined signals. It is becoming quite popular.

Motorola and Intel have independently developed incompatible bus systems. Their intention seems to be to lock the user into further proprietary equipment. The Motorola Exorciser bus supports the M6800 development system for microprocessor control systems. The Intel

Multibus supports the Intellec development system; it has been standardised as IEEE-796.

The original PDP-11 UNIBUS (see Section 4.4.1) has been replaced by the LSI-11 Bus with multiplexed data and address lines. The principal features of the UNIBUS handshake are retained.

9.1.2 Purpose-built systems

Standard off-the-shelf modules must be designed to suit a range of applications, and must involve some compromise. Expedience favours ready-made equipment, but it is often the case that such equipment cannot achieve the performance required, or that more elaborate (and expensive) equipment cannot be justified for an exploratory or short project. In these cases it may be necessary or worthwhile to design and develop a special system.

Conventional techniques for data-collection and control interfaces were described in Chapter 7. The concepts of DAC, ADC, timer/counters, VFC and FVC were all discussed in sufficient detail for readers to assess the suitability of each for their own applications.

Chapter 7 also addressed the speed limitations imposed by programmed, interrupt-driven or DMA interaction schemes. The reader should be able to assess the feasibility and limitations of any interface/interaction scheme for conventional applications. The next section will describe some examples of such systems.

9.2 An interrupt-driven data-collection/control system

This section describes a system in an Apple microcomputer, based on the content of Chapter 7. It was built to study the behaviour of a novel electrostatic lens element for ion and electron beams. Electron (and ion) lenses are best known as the focussing elements in electron microscopes, but they have many other applications in cathode ray oscilloscopes, television picture tubes and ion microprobes. The lenses used in these instruments are quite well understood, but their performance is nowhere near as well developed as that of the lenses used in light-optics. The individual lenses in cameras, microscopes etc. are far from perfect, but they are combined into a system such that the individual imperfections or *aberrations* of each element are mutually corrected or compensated. The theory of light-optics has been developed to the stage where it is possible to design and build lens systems which are virtually perfect; their performance is limited only by the wave nature of light, that is, by the diffraction limit of the lens aperture.

The refracting or reflecting surfaces of lenses for light-optics are sharply defined; the theoretical analysis is therefore based on discontinuous media. The electric or magnetic fields used in electron or ion lenses change continuously through the lens space. Light-optic theory is grafted uncomfortably onto these fields and, as a result, lenses for electrons and ions are nowhere near as well developed.

Aberrations are only tolerable very close to the optical axis and at very small angles to it; electron microscope lenses, for example, can only operate at very small apertures and with very small angular acceptance.

There have been very few successful attempts to apply the principles of aberration correction to ion and electron lenses. The reasons for this can be put crudely, as the unsuitability of conventional axially-symmetric lens geometries for mutual aberration compensation. In a search for suitable new lens elements, we have been exploring bounded *planar* elements with no axial symmetry and with only one or two planes of reflection symmetry. These look very promising, but their novelty makes their analysis by conventional theories rather tedious. To avoid the need for a new, precise theoretical model, we surveyed their performance in the first instance, with a semiempirical technique. Their general behaviour was assessed with a crude but quick approximate model to explore their suitability for aberration compensation. Their precise behaviour was then determined empirically, by measuring the deflection of carefully controlled, thin ribbons of beam fired through the lens. Fig. 9.1 shows the general experiment layout.

The planar lens element or system is mounted behind a small defining slit with a gap of about 30 μm; this slit can be moved by a stepper motor (see Box 9.1), across a collimated ion beam coming from an ion accelerator. A finely collimated beam of ions at a known position and angle, passes through this slit into the lens system.

Downstream from the lens, a detector slit with a gap width of about 10 μm can be scanned across the beam by a second stepper motor. An ion detector, mounted behind this slit, monitors the intensity of the ion beam at the detector slit's position. When the slit is moved, the intensity profile and hence the beam deflection can be measured.

Both stepper motors are driven from a straightforward interface using two Philips SAA1027 stepper motor control chips and simple support electronics to interface them through three I/O ports onto the Apple busses. The card allows either motor to be moved 1–127 steps in either direction by a single program command. A status register in the interface indicates when the motion is finished.

The ion detector is a Phillips Channel Electron Multiplier (CEM). It consists of a coiled glass tube open at one end, and with a high-resistance coating on its inner surface; a potential difference of a few thousand volts is applied along the length of the tube. When an energetic electron or ion strikes the inside of the tube near its open end it releases a few secondary electrons which accelerate down the potential gradient. When they collide with the walls they generate a shower of secondaries. If the potential gradient is properly set, an electron avalanche results with a pulse gain of about 10^5. This generates a detectable charge pulse from the collision of a single ion or electron at the mouth. (The CEM operates rather like a photomultiplier tube, but the simple electron optics of the CEM generates a less-well-defined pulse.)

The charge pulse is passed through a system of standard NIM modules (Nuclear Instrumentation Manufacturers). These are used extensively for nuclear instrumentation; they consist of a range of compatible modules for detecting, shaping, amplifying and processing the electronic pulses generated by particle detectors. The set-up used here consists of a high-voltage supply for the CEM, a preamplifier placed close to the detector (to

Figure 9.1 (a) Experimental arrangement for ion lens performance assessment, (b) flow chart of the interrupt routine to collect the raw data.

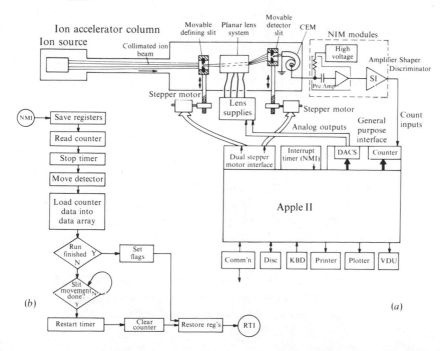

convert its high-impedance output to a low-impedance signal, less susceptible to noise), a shaper/amplifier which converts the charge pulse into a smooth quasi-gaussian pulse whose height is proportional to the integrated intensity of the charge, and a discriminator which generates a fixed output pulse when the height of the shaped pulse is between presettable high and low levels. The discriminator output pulse is TTL-compatible and is passed to a 16-bit counter/timer in the general-purpose interface card. (This card was built for general teaching and research work; it provides a timer/counter module, ADC/DAC functions and digital I/O ports.)

When the ion detection system is properly adjusted, the ion flux on the detector can be measured by counting pulses. Event counting requires the time interval to be controlled precisely. A simple programmable timer is used to perform this timing; it generates a NMI interrupt at programmable intervals (from one millisecond to one day). It can be left to run continuously, generating interrupts at the programmed interval, or can be disabled so that it starts timing from an instant determined by program commands.

The electrostatic potentials applied to the lens elements are generated from HT power supplies controlled by DACs within the general-purpose interface.

The software operates at two levels. A BASIC foreground program handles the setting of lens voltages, movement of the defining or detector slit, setting of the steps and range of the slit motion, controlling the time interval between slit movement, setting display parameters etc. This is all menu-driven and allows easy manipulation of all experimental parameters.

The experiment is started or stopped (from the menu table), by enabling or disabling the interrupt timer. The interrupt service routine is shown as a flow chart in Fig. 9.1(*b*). After saving the registers, the counter is read and the interrupt timer is stopped. The stepper motor is then instructed to move the required distance for the next reading and the collected data is added to the corresponding channel of the data array; the channel pointer is then incremented. If the run is not finished, the stepper motor interface is watched until the motor stops, when the timer is restarted and the counter is cleared. The routine thus assures that counts are accumulated only during the time specified by the timer (with a small known error due to software overhead).

In our experimental set-up, the total beam current is about 10^{-6} A, but the current reaching the detector is less than 10^{-14} A (10^5 ions per second). It is necessary to count for about $\frac{1}{2}$ s to achieve good statistics. (The system is limited by the NIM modules to about 100 kHz, but the interface electronics can handle up to 10 MHz.)

The system collects data into an array of up to 1024 count-channels as a function of the detector slit position. The data is tallied by the interrupt routine directly into a BASIC integer array, easily accessed by the foreground BASIC program. This allows convenient BASIC commands, in immediate or deferred mode, to monitor the progress of the experiment, and to test status and conditions during the experiment.

The interrupt routine only requires a very small portion of the processor's time. Since it is initiated by a NMI signal, it gets rapid access to the processor. The BASIC foreground program is left running to monitor progress, to display the data being collected and to respond to user commands. It can also decide when to stop a scan, when to change the defining slit position, when to change lens voltages etc.

This application demonstrates the principles of multi-channel data collection; it is an example of *Multi-Channel Scaling* (*MCS*) where the data is counted (scaled) into an array of channels. The use of a counter in the interface allows data to be collected by a simple system like this at up to 10 MHz. Another application of the general scheme is the *Multi-Channel Analyser* (*MCA*) arrangement wherein different types of events (different pulse magnitudes, for example) arc classified and tallied separately into an array.

Box 9.1
Mechanical actuators and position encoders

Instrumentation for mechanical systems commonly requires actuators to move system components, and detectors to determine their positions. Servo-motors combine both functions (Fig. 9.2) by having an internal position transducer which generates an analog signal. This is continuously compared with an analog reference signal and the difference between them (the error signal) is used to drive the motor such that the error is reduced. Servo-motors and servo-actuators are available as very small, cheap units intended for model aeroplanes etc. through to very powerful and expensive machines for precise linear or rotational positioning.

Within a computer-based system, the error detection and control feedback functions can be performed by optimal control software. The program can devise the best strategy for moving the actuator into a new position in the fastest time or with minimum energy consumption etc.

Precise control requires transducers that provide accurate, high-resolution information about the position of the mechanisms. Precision, digital, position encoders suitable for computer interfacing are widely available from many suppliers. There are two broad classes, absolute or incremental. Both types are available with very high resolution; rotary encoders can resolve 2^{16} divisions in a cycle, linear encoders can resolve microns.

Absolute encoders provide a data word corresponding to the distance that the detector has moved from its origin, incremental encoders give an output pulse for every quantum of motion of the detector (with some means of indicating the direction). An incremental encoder only requires an up/down pulse counter to mimic an absolute encoder.

A simple, low-cost, incremental encoder can be made with a simple chopper disc and two photodetector assemblies. If the outputs from the two detectors are arranged to be 90° out of phase, they can be used directly as the logic inputs for a 74LS193 counter array (see Fig. 9.3).

Mechanical actuators can be made from any rotary or linear motor that can be reversed. Permanent magnet d.c. motors only need a bipolar amplifier driven from a DAC interface to provide full control over acceleration, speed and position. (See the motor controller chips such as National LM1014.) A.C. motors require a bit more support. Single- and three-phase motors can be reversed by changing the phase of their windings. This can be done for small motors by an analog or digitally controlled phase-shift circuit; large motors usually require a contactor or TRIAC controller system. Hydraulic and pneumatic actuators can be

Figure 9.2   Servo-motor configuration.

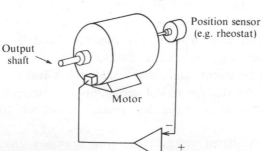

operated and controlled by readily available solenoid valves powered from relays or current amplifiers.

Stepper motors represent a simple actuator with inherent incremental encoding. They consist of a permanent magnet rotor with alternate N and S poles. This is mounted within an electromagnetic stator which allows a matching pattern of magnetic poles to be moved one step at a time in either direction. Provided that the magnetic coupling force always exceeds the load, the rotor faithfully follows the moving stator field. Small stepper motors are now used extensively in printers and plotters; they usually have four-phase windings which can be driven with monopolar logic signals. (The newer NEC printers use a linear stepper motor for the print-head motion.) Integrated circuits such as the Philips SAA1027 allow stepper motors to be easily interfaced to microcomputers.

Quartz piezo-electric devices are very useful for small-scale actuators and motion detectors. (They are used to control the position of the heads in hard-disc memory systems and the new laser disc players.) The *Inchworm* device built by Burleigh Instruments of New York is a clever but expensive application of piezo-electric actuators (Fig. 9.4). It is made from three bonded piezo-electric cylinders; the two end cylinders can be expanded or contracted radially to release or clamp a smooth metal shaft; the centre cylinder can be expanded or contracted axially. By programming the actuation of all three cylinders, it is possible to move the shaft in either direction by small steps, down to about 10 nm. The makers claim that the

Figure 9.3   A simple optical incremental position encoder using 74LS193 chips.

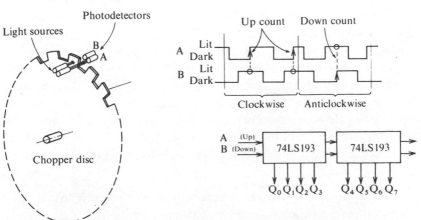

Figure 9.4   The *Inchworm* actuator

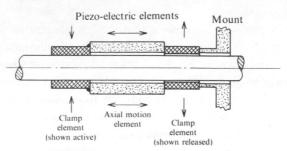

travel can be determined reliably and repeatably, from the step size and step count, but they also offer an optional optical position sensor.

9.3   DMA systems

The interrupt-driven application described in Section 9.2 is well within the speed and memory capability of a small microcomputer. There are many applications where the processor's program execution rate is inadequate to handle the data input; there are also many applications where the memory space is too small for the database to be collected. This section describes two examples which handle high data rates by DMA interaction within the friendly framework of a simple 8-bit microcomputer.

The first example is a system to collect information from the image data of a Scanning Electron Microscope (SEM) using a Robinson detector. The backscattered electron current from the detector is proportional to the atomic mass number of the target area illuminated by the electron beam. The system is required to generate and display in real-time, a histogram of the relative abundance of chemical compounds in the SEM field of view, and to allow real-time interaction. It is used for assaying ores. The histogram is generated by repeated sampling of the secondary electron current; each sample is digitised by an ADC to yield a corresponding channel number for that sample, and the corresponding element of an integer array is incremented. This generates a multi-channel spectrum representing the occurrences of each channel's assigned analog range. This is an MCA spectrum of the analog variable. Being essentially a counting system, it is capable of handling very large relative variations between channels.

An MCA algorithm can be implemented quite easily in an 8-bit microcomputer by the following sequence of operations.

(1) The channel number must be augmented to define the memory location assigned to the channel.
(2) The contents of that memory location must be fetched, incremented and returned to memory.
(3) If Step (2) resulted in an arithmetic overflow, either an increment sequence must be performed on the memory word assigned to the next higher byte of the channel, or some form of overflow flag must be given.

When the algorithm is implemented in software, the minimum processor time for each sample is about 30 μs; overflow handling doubles the time required. Interrupt access adds to the overheads. If this data rate is adequate, an MCA system can be implemented in the same manner as the previous example. Our example requires samples to be taken at about 50 μs intervals. This is just within the range of program control, but the system must also generate a rather complicated colour display in real-time and analyse the data to display relative intensities for several compounds of interest. The OKI (Z80) computer used for the system was chosen for its good colour graphics capability. It is barely able to keep up with the data display and analysis, with almost no time left for data collection. DMA interaction is essential.

The MCA spectrum requires 1024 channels of 16 bits each. The memory array consists of 2048 memory locations; the 10-bit channel pointer is located in the computer's memory by a 5-bit array pointer programmed via a control/status port in the Z80 I/O address space. The l.s.b. of the augmented 16-bit address points to one of the two bytes corresponding to each channel (see Fig. 9.5). The top 3 bits of the port word are used as flags for RUN/STOP and INCREMENT/DECREMENT control functions and data overflow status.

At the start of each sample period, the ADC carries out a conversion; when complete, the DMA logic requests bus access via the BUS REQUEST line. The DMA logic then deposits the augmented address (with A_0 set low) on the address bus and requests data from the specified location, latching it into an 8-bit counter. The counter is then incremented (or decremented) by one; an overflow (or underflow) result is latched for later testing. The modified data is written back into memory during the next bus cycle. If an overflow(/underflow) occurred, the cycle is repeated with the A_0 address line set high to increment (or decrement) the high-order channel byte.

A single cycle is executed in two Z80 bus cycles (500 ns); an overflow condition requires two more bus cycles. The average access time is thus 502 ns per sample, or just over 1% of the processor time at the 20 kHz sample rate. The data-collection process is virtually invisible to the program.

The second DMA example involves an Apple used as a low-angle X-ray diffraction analysis system.

When a polycrystalline powder sample is illuminated by a finely collimated X-ray beam it generates a diffraction pattern consisting of concentric rings. Conventional photographic techniques require very long exposures to detect faint rings generated by trace compounds, but these tend to be swamped by the strong rings of dominant compounds. Pulse counting techniques, however, allow a wide dynamic range and separate the strong and weak signals.

A commercial, linear-array, position-sensitive X-ray detector is used to collect data in a diagonal line across the ring pattern. This detector consists of a gas-filled chamber with a very fine wire held at a high voltage, mounted close to an array of anodes connected as a lumped-parameter delay line (see Fig. 9.6). When an X-ray passes through the chamber it ionises the background gas and initiates a brief arc (rapidly suppressed by components of the background gas) in the high-field region near the wire. The short burst of charge reaching the nearest anode propagates as two pulses in either direction along the delay line. The difference between the arrival times at each end carries information about the source of the pulse. Commercial NIM pulse processing modules are used to convert the time difference into a short pulse whose amplitude is proportional to the time difference (a Time to Pulse-Height Convertor (TPHC)).

Figure 9.5   MCA system operating under DMA.

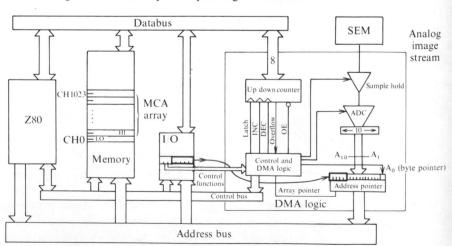

The output pulse is detected and digitised by a sample/hold ADC circuit and is used to produce an MCA spectrum of X-ray counts (integrated intensity) as a function of position along the detector. The TPHC system can detect pulse pairs up to about 100 kHz. Four convertors can be used on the system and a maximum pulse rate of about 400 kHz must be handled. This allows only about two microseconds per pulse and can only just be handled by DMA interaction in an Apple, using all the available bus transaction time.

The DMA access is organised in a similar manner to that of the previous example, each spectrum consists of 1024 16-bit channels. Subsystems for each of the four detectors interact via a daisy-chain DMA request structure. As pulses are detected and converted, the subsystems queue up for orderly access to the DMA logic.

The system is built in a four-card rack; priority on the daisy-chain is rack-position dependent so that the priority level of each subsystem can be manipulated by moving the cards. The location of each MCA array in memory is defined independently by control/status registers in each subsystem; these are usually programmed so that the four spectra correspond to a 4×1024 Applesoft integer array. This makes it very convenient to monitor the data-collection process from simple BASIC programs. Note, however, that when the DMA logic is very busy the processor almost stalls!

9.4 Extending performance beyond the microcomputer's capability

The two DMA examples described in the previous section provide a considerable improvement in performance when compared with simple

Figure 9.6   A gas-discharge, position-sensitive X-ray detector.

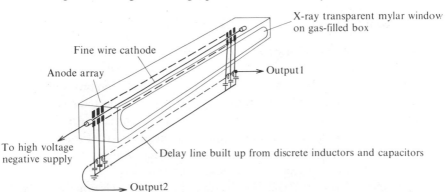

programmed or interrupt driven interaction. They also demonstrate some of the problems caused by the architecture of the computer. The four major problems are:

(1) The transaction cycle of the computer limits the data-collection rate and prohibits read/modify/write memory cycles.

(2) Large data words must be broken up into the width of the databus and handled serially. This extends the time needed for data manipulation.

(3) If the data-collection process is very intense, DMA access to the microcomputer busses causes a serious loss of processor time. (Minicomputers handle this better, where bus cycles can be stolen; see Section 7.8.)

(4) The limited address space available in the host computer restricts the space available for DMA use; large arrays restrict the program space seriously.

The rest of this section describes some applications which avoid all of these problems. The solution depends on designing an external bus architecture which suits the needs of the data-collection system, but is able to communicate smoothly with the host processor.

9.4.1   DPM systems

The busy-bus syndrome of the previous example can only be eliminated by freeing the computer busses from any involvement in the data-collection process. The DPM architecture described in Section 7.9 is able to achieve this.

With some care in its implementation, this architecture can be made not only to move the entire data-collection process off the computer bus, but also to provide a very fast data-collection system freed of all the time and space restraints of the computer's bus structure.

The next example describes a *clipped digital correlator* organised as a fast external data-collection system and interfaced into an Apple microcomputer as if it were a block of DPM.

Signal correlators are used in many fields such as Astronomy and laser light-scattering to filter faint repetitive or coherent signals out of very large broad-band noise. The tuned circuit in a radio or television receiver filters a narrow frequency band out of the broad-band noise collected by an aerial; it is a bit like a single-channel correlator. A multi-channel auto-correlator does much the same for a range of signal frequencies by generating an array of correlation coefficients between the instantaneous signal intensity and its intensity over a range of previous instants.

In a clipped, digital correlator, the time varying signal is sampled at fixed intervals and the instantaneous value at that time is *clipped* to a binary logic level by being compared with a reference value. This clipped signal is passed into a binary shift register and shifted at each time interval. The auto-correlation function is generated by multiplying (**AND**ing) the present clipped level with every output of the shift register, and accumulating (adding) the correlation terms (**1** or **0**) into an array of counter channels. Fig. 9.7 depicts the general layout of the logic.

In radio-astronomy applications, the data input is a broad-band radio signal collected by an antenna. The intensity of this signal as a function of frequency is of considerable interest, and can be extracted by auto-correlation in a clipped digital correlator. Auto-correlation is performed by the **not-EOR** logic function to generate a spectrum corresponding to signal phase correlation. This is intimately related to the signal's power spectrum over the range of corresponding frequencies. A simple demonstration system was needed some time ago to demonstrate the principles for undergraduate courses. It was designed to fit into an Apple II and to accommodate up to 1024 counter channels. It operates at a maximum sample rate of 7 MHz, thus all logic operations must be performed in parallel by dedicated hardware. A block layout of the system components is shown in Fig. 9.8; it is an adaptation of DPM architecture. The array of latches (the storage registers of the 74LS590 counters) appears as simple memory in the I/O SELECT area of the interface slot used. The latches are read-only and cannot be written by the computer, but a common CLEAR signal, operated through the I/O STATUS/CONTROL port in the peripheral logic allows the entire array to be cleared to zero counts. The array is configured in modular blocks of 128 16-bit counters, mapped onto

Figure 9.7   The clipped digital auto-correlator layout.

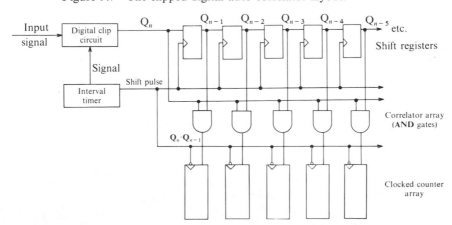

the Apple interface slot I/O SELECT page as 256 bytes. When the processor reads the registers, the address select logic simply decodes the address bus contents and enables the required 74LS590 to deposit its data onto the databus. Modular blocks are selected by decode logic driven from the I/O CONTROL port.

The DPM scheme must provide for contention between the processor and the peripheral logic access to the shared register/memory block; simultaneous use of the block by both subsystems would cause errors. The ideal situation is where both have *transparent access* to the memory block. This means that neither subsystem obstructs or delays the other. This is implemented easily in the system shown in Fig. 9.8, by preventing the register array from being latched while the processor is accessing the memory block. The I/O SELECT signal handles this perfectly.

The sample rate of the incoming signal is determined by a prescaler controlled by a simple output port in the Apple I/O space. The whole process is further controlled by a second I/O port which allows the process to be turned off or on at will and to change the logic correlation function. The clipper level can also be controlled by software.

Figure 9.8   Block/schematic layout of a clipped digital correlator system fitted to an Apple microcomputer. Only part of one module is shown. 'CLEAR' logic is not shown.

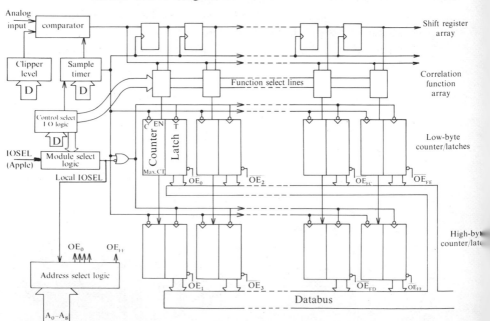

Auto-correlation values for typical astronomical signals are very low. The process must be left to run for a long time before a significant variation between channels (of the order 10^{-4}–10^{-6}) appears. Although these very small variations would seem to require a word of about 20 bits, they can be detected with a 16-bit counter array by pretending that the counters have more bits; the correlation differences appear eventually in the lower 16 bits. One channel of each module array is provided with a 24-bit counter array; the extra data is used to evaluate the absolute correlation coefficients. This byte is available through a DEVICE SELECT input port.

9.4.2 Autonomous peripheral data-collection systems

The performance of a conventional peripheral interface is limited by the data width and the transaction cycle time of the computer. The previous example shows how the limitation can be broken by an autonomous data-collection system. The data-collection hardware has grown from a simple interface into a stand-alone system, linked to the computer by shared access to its data array, but controlled and monitored by the computer software. Freed from the need to handle the data-collection process, the computer can use its time to monitor and manipulate the database at a high level, and to make decisions about strategies.

The previous example created a very fast external system by using TTL counter/latches and making them appear to be ROM. In general, the external memory must be R/W. If it need not be very big, a fast peripheral system can be built from bipolar or ECL memory, but large memory arrays using either of these need a lot of power, space and money.

The most economic memory, with the lowest (bit cost) × (access time), is the large array DRAM chip. These have access times down to 150 ns; a R/W cycle can be performed in something like 300 ns. When used outside of the computer, their ability to perform a read/modify/write operation in a single sequence (about 350 ns) can be used. Their main disadvantage is the need to provide refresh cycles.

Autonomous peripheral systems require much more design and development than a conventional interface, hence they should only be considered when the need for them has been proven. The data-collection example described in Section 9.2 was actually patched together from standard peripherals developed for earlier projects. However, when the research results looked very promising, a new dedicated system was built. This replaced the program-driven interaction, collecting data from one channel at a time, with a much faster technique which collects data simultaneously for all channels.

The experimental set-up had to be modified; the previous scanning slit defining the ion-beam was replaced by a fixed array of fine slits. The multiple fine beams, after passing through the lens, now fall on a scintillator screen. The light produced by the screen is collected and focussed on an array of light-sensitive detectors, a linear-array CCD (Charge Coupled Device).

Rather than bore you with more details of this revamped system, the final example will describe a system which has been developed from it, and which supports two-dimensional-array CCD devices. This example also shows the external data-collection system organised with a large data-word to suit the data-collection process, yet matched to the computer databus width when accessed from the computer.

The CCD chip is a fine example of advanced microelectronic design. Fig. 9.9 shows the (simplified) layout of the GEC P8600 chip that we use. It has 385 electron *channels* formed by selective doping of the silicon substrate, and separated by *channel stop* regions (where the electrons cannot penetrate). Repeated sequences of insulated polysilicon conducting stripes are laid down at right angles to the channels and are connected at their ends to create two separately controlled 288-cycle three-phase electrostatic linear motors. Light falling on the detector creates electron–hole pairs; the free electrons generated in the channels are attracted to the nearest motor

Figure 9.9   The layout of the GEC CCD *frame-transfer* video camera chip (greatly simplified and only a small section shown).

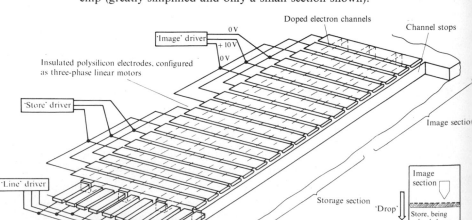

phase line with a positive potential (with respect to the substrate). The detector thus integrates the light intensity independently for all 385×288 regions defined by the channels and the motor array. At the end of the integration time, both linear motors are run together at about 1 MHz to sweep the entire charge distribution down into the storage section. The image section is then left to continue collecting charge, while the image charge array (now in the storage section) is shifted by its separate motor, one row at a time, into the transverse linear motor. This propels an entire charge row, one packet at a time, into a charge-sensitive detector diode. The entire stored picture is thus unloaded, in normal video picture format, as a frame composed of individual pixel charge packets.

These chips represent a powerful detection system for large data arrays that can be coupled into the CCD array as a pattern of light intensity. More than 100 000 integrating detectors are operating simultaneously, and the system allows simple logic extraction of the information through a single processing port.

Our applications for the two-dimensional-array processor system are in astronomy, spectrometry, low-angle X-ray diffraction analysis and undergraduate laboratory work. All of these applications are rather demanding in that the data to be collected has a contrast ratio of the order of more than 10 000/1, and the lowest-level signals are generally below the thermal noise level of the CCD array. This requires background noise to be reduced as far as possible by cooling the detector with a Peltier cell or with liquid nitrogen, and using signal/noise enhancement techniques to extract the real signal from residual noise.

Fortunately, however, all of our applications are at low light levels where the CCD need not be run at normal video frame-scan rates. Slow-scan techniques allow the use of economical 64K DRAM memory chips.

The system's memory is organised as 64K words of 32 bits. This is assigned to a 256×256 area of the CCD chip. This wastes about 40 % of the detector, but greatly simplifies the system logic design in the first case. (Subsequent development of the system will extend the memory span.)

Twenty-four bits of the memory word are used for data accumulation and signal/noise enhancement. Conversion of the analog output from the CCD is done by an 8-bit flash convertor. The dedicated logic of the system allows the memory array to be driven in the efficient read/modify/write mode with a cycle time of about 400 ns.

Fig. 9.10 shows a block layout of the system components. The output lines of the memory array are fed to a hardwired 24-bit add/subtract logic block which allows the new 8-bit data coming from the ADC to be added or subtracted. The 24-bit output of this block is fed back to the memory input lines.

The maximum achievable scan rate is about 20 frames per second; this is too fast for most of our applications, which must be controllable over a range of about 20–0.01 frames per second. This is achieved by running the store section of the CCD, and the memory, continuously at full speed, but without writing back into memory. This keeps the store section 'clean' and provides an automatic refresh of the DRAM memory. A frame-scan counter, programmed from the computer, tallies the number of these 'dummy' scans. At the programmed scan count (i.e. time), the charge pattern in the image section is transferred into the storage section and, during the subsequent frame unload, the memory is written with the add/subtract output data.

During data acquisition, the memory is fully occupied and cannot be accessed; however, during the scan flyback time and the unused part of the line, the host computer can gain free access. Simple contention logic handles this easily.

The advantage of the large array of detectors is gained at the expense of calibration problems. The gain of individual detectors in the CCD array varies by about $\pm 25\%$. For serious data-collection work, it is necessary to calibrate each detector and to compensate the accumulated data array. The spare 8 bits of the 32-bit memory word are used to carry out real-time gain compensation. Before the experimental data is collected, the system is run for some time with even illumination. The resulting data array is processed and loaded into the extra memory as the sensitivity matrix of the CCD

Figure 9.10   Block layout of the autonomous, peripheral system to support an area-sensitive CCD.

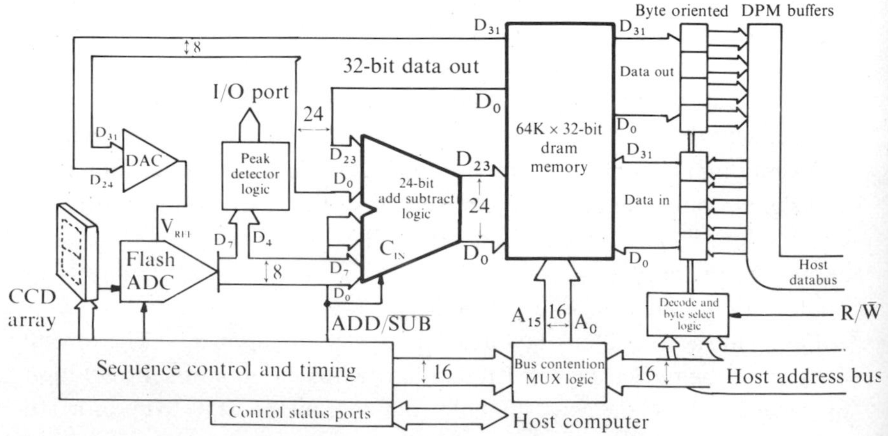

array. When data is being collected subsequently, the sensitivity array is automatically scanned and the sensitivity data for the element about to be converted is sent to a fast DAC. This provides the reference voltage for the ADC.

The top 4 bits of the ADC output are passed to a 4-bit comparator. This is arranged such that it latches the input into a 4-bit register if it is greater than the register's current value. The register can be read and cleared from the host computer. This simple logic block is a peak value detector; software in the host computer can monitor the conversion and can control the integration time to optimise the data-collection efficiency.

All of these features allow the host computer, free of any involvement in the routine task of data collection, to operate and control the system in many modes. The computer is not able to handle the maximum data-collection rate, but it can *monitor* the entire array or some nominated section of it, and can adjust the system state as required, to optimise aspects of its overall operation.

Signal recovery from background noise can be achieved with extra hardware to modulate the real signal whilst leaving the noise constant. If the system is set to add the data while signal and noise are present, but to subtract the data for the same time while the signal is absent, the real signal will be enhanced from the noise. In astronomy work, this is achieved by regularly interposing a shutter in the optic system, or by moving the field of view to a blank area. In low-angle, X-ray diffraction work, the sample is moved in and out of the finely collimated X-ray beam.

Long-term integration of the signal, or the noise reduction schemes outlined above, allow the system to record intensities below the resolution limit corresponding to the l.s.b. of the ADC. The Poisson statistics of random events assures that if one waits long enough a burst of events will arrive within one integration period, sufficient to appear above the detection threshold. The limiting case is where the expected signal is overwhelmed by random fluctuation of the noise signal. Extensive literature is available on this subject of signal recovery techniques.

The system can be fitted into any suitable 8-bit or 16-bit computer. The dual-ported memory buffer can be easily arranged to communicate 8-bit or 16-bit data with the host. Our systems are fitted to NEC APC 16-bit (8086) microcomputers. These have a good open structure and have enough memory space for the system memory to map directly into the machine as 256 Kbytes of normal memory. When the external system is not using the memory, it appears to the microprocessor as normal expansion memory.

If the processor attempts to access the external memory during data acquisition, it is forced into WAIT states; this time loss can be avoided by making the processor first test the external status word to check if the external memory is in use. During the inactive section of each line scan, and during line and frame change, the processor can steal data cycles from the external memory.

This kind of trade-off must be accepted when a system stretches the capability of its components; the memory is being used close to its limit to collect the CCD data, there is just no time left for processor access. FIFO registers could be installed between the two systems to allow processor data requests to be queued up and serviced when the external system is free.

The system architecture could support data collection at TV scan rates if DRAM memory were replaced with fast ECL or TTL memory.

9.5   Conclusion

The principal aim of this book has been to reveal the versatility and power of the present generation of microcomputers in data-collection and control applications. Such applications are not in the main-line of computer usage and hence are not very well known or taught. It is hoped that this book will allow those involved in Science and Engineering to realise the full potential of the excellent and cheap little computers which are now freely available.

No prior knowledge of computers or electronics has been presumed, yet the intention has been to bring the reader to a full understanding of its internal operation. This has allowed the contents of Chapter 8 to give the reader a broad insight into the operation of some common computer peripheral systems and into some of the concepts of computer system architecture.

It has not been possible to cover all aspects of this very extensive field in a single book. The last two chapters have perhaps attempted to cover too much but, even if they are a bit thin, they provide a friendly entry into the extensive and ever-growing literature.

There are many glaring gaps in the content of the book. No attempt has been made to describe the extensive range of primary transducers now available for the data-collection 'front ends', neither has there been any discussion of the many, elegant, mathematical techniques which can be used to process, reduce, transform and otherwise analyse large data arrays which have been collected. Both of these subjects are covered very well in many other books. We have elected to concentrate on hardware techniques for data collection and control with a microcomputer.

It is hoped that the content of Chapters 7 and 9 may encourage some brave souls to go boldly forth into new fields, and to 'have-a-go' at designing and building their own data-collection interfaces. If this is your bent, do not start with complicated systems. Start instead with something very simple such as an 8-bit digital I/O; the many tricks involved in commissioning a digital system should be learnt on the simplest possible projects.

One never expects a thousand-line program to run straight away, it must be written in separate testable modules. Bear in mind that digital hardware is very like a complex program, and must be developed in the same manner. Design it in such a way that simple modules of the circuitry can be individually tested and commissioned. If this procedure is followed rigorously, the commissioning of a complete system is surprisingly hassle-free and is very rewarding; as modules are commissioned, one can almost feel the entire system coming alive gradually as its intelligence grows. One realises the motivation that surely must have driven Count Frankenstein; I hope and trust that none of your creations will turn out like his. Good luck!

References and further reading

1 Integrated circuit manuals from all of the major manufacturers such as Fairchild, Motorola, National, NEC, Intel, Datel, Analog devices, Burr-Brown, Hewlett-Packard etc.

2 *Logic Designer's Guidebook*, E. A. Parr, McGraw-Hill, New York (1984)

3 'The IEEE Standard for the S100 Bus', M. Gartez, *Byte*, Feb. 1983, pp. 272–97

4 *STD Technical Manual and Product Catalog*, STD/Prolog Corp. (USA)

5 *Data Communications for Microcomputers*, E. A. Nichols, J. C. Nichols & K. R. Musson, McGraw-Hill, New York (1982)

6 *Interfacing to Microprocessors*, J. C. Cluley, McGraw-Hill, New York (1983)

7 *Basic Principles and Practice of Microprocessors*, D. E. Heffer *et al.*, Edward Arnold, London (1981)

8 *Microprocessor Support Chips*, T. J. Byers, McGraw-Hill, New York (1984)

9 'Interfacing for Real-Time Control', R. M. Gennet *et al.*, *Byte*, Apr. 1984 *et seq.*
    (The *Byte* issue of Apr. 1984 has several articles on real-time measurement and control.)

Electronic trade journals, and magazines such as Byte, *contain information on new products and reference books. Device manufacturers, or their distributors, provide data sheets for new devices on request.*

Index and glossary